Seaweeds of Long Island Sound

Second Edition

by Margaret "Peg" Stewart Van Patten
with Foreword by Dr. Charles Yarish

Acknowledgments

Many thanks to the following inspiring experts who contributed immensely as technical advisors, provided specimens and images, and also reviewed this guide:

James F. Foertch, Millstone Environmental Laboratory
John T. Swenarton, Millstone Environmental Laboratory
Sally L. Taylor, Professor Emerita of Botany, Connecticut College
Charles Yarish, University of Connecticut at Stamford

In addition, the following individuals contributed time, knowledge, images, suggestions or general support for the project: Edward C. Monahan and Sylvain De Guise, University of Connecticut; Sally Taylor, Connecticut College Arboretum; Shelley Cudiner and Nancy Gillies, librarians at the University of Connecticut's Jeremy Richard Library; and Dr. James Sears, UMass Dartmouth. The author is grateful to the University of Connecticut Avery Point campus for use of microscopes. Photos by Courtnay Janiak, Charles Yarish, Jim Foertch, John Swenarton, Jim Sears, and others added greatly to this guide's appeal.

Funding for this project was provided by the United States Environmental Protection Agency's Long Island Sound Study, Connecticut Sea Grant, and the Connecticut College Arboretum. The author and collaborators thank these generous sponsors.

About the Author

Margaret ("Peg") Van Patten, B.A., M.S., is Communications Director for Connecticut Sea Grant, at the University of Connecticut at Avery Point. She is a graduate of both Connecticut College and the University of Connecticut, with degrees in Human Ecology and Marine Sciences. She has done research on the reproduction of *Laminaria* in Long Island Sound and greatly enjoys collecting seaweed. Peg hopes that this guide will spark an interest in the algae as well as Long Island Sound for many people, including her children and grandchildren, when they visit Long Island Sound.

VanPatten, Margaret (Peg) Stewart Van Patten
Seaweeds of Long Island Sound. Includes index and bibliography. Second edition.
1. Marine algae–Atlantic coast (Long Island Sound, (N.Y. and Conn. United States) Identification.
2. Marine flora–Atlantic coast (N.Y. and Conn. United States) identification.

ISBN 1-878301-09-8

Contents

Foreword

Long Island Sound, at 41° N, 72-73° W, has a long coastline–about 110 miles if a straight line is drawn in the center from end to end. It extends from the shores of New York City in the west, to Fishers Island in the east. On its northern shore is the coastline of Connecticut with its plethora of rocky habitats, and on its south is the sediment coast of Long Island, New York. It has many offshore islands, a diversity of coastal habitats, and a wide range of temperatures, encouraging the growth of a diverse flora of seaweed species, despite being an urbanized estuary that is impacted by more than 15 million people. Shoreline walkers, beachcombers, and those who enjoy working, living, or just playing on the shores of The Sound, have often expressed interest for a current, non-technical field guide to the common seaweeds of the region. The popular *Seaweeds of the Connecticut Shore - A Wader's Guide* (S. L. Taylor and M. Villalard, Connecticut College Arboretum) and *An Annotated Checklist of Connecticut Seaweeds* (C.W. Schneider, M. M. Suyemoto and C. Yarish, Connecticut Geological and Natural History Survey. Bulletin 108) together filled the niche for many years, but now they are both out-of-print and outdated due to recent changes in seaweed classification.

Whether you are a nature enthusiast, a teacher, a student or a scientist, this guide is intended to stimulate your curiosity and assist you with the identification of these very unique and beautiful photosynthetic marine organisms inhabiting the shores and shallow waters of Long Island Sound. Every effort has been made to incorporate habitat color images of these seaweeds, and when possible, photomicrographs have been included. Descriptions of the seaweeds include external form and structure, texture, color, and habitat in Long Island Sound.

It has been a pleasure assisting Peg with this project. We hope this guide will fill the user with the excitement experienced by those who know the seaweeds well. Finally, we must add that without the collaboration between the various campuses of The University of Connecticut, the Connecticut Sea Grant College Program, The U.S. EPA Long Island Sound Study, and the Connecticut College Arboretum, this guide would not have been possible. Such partnerships are the way to make more and better information available to larger audiences.

Dr. Charles Yarish

Dr. Yarish is Professor of Ecology and Evolutionary Biology, and Marine Sciences, at the University of Connecticut, a member of the Northeast Algal Society, and former President of the Phycological Society of America.

Introduction

There is a truly amazing variety of seaweed, known to scientists as macroalgae, in the waters of Long Island Sound. Unfortunately, many shoreline visitors know seaweed only as "slimy stuff" and never experience the natural symmetry and innate beauty of the large algae. Nor do they appreciate the ecological and economic importance of these organisms. Until now, there have not been many publications available to help the public appreciate and learn about the seaweeds. This guide is intended for the curious beachcomber, rather than the biologist, and hopefully will improve the reputation of seaweed in our region.

The Sound has such a rich variety of algae because its variety of habitats, large temperature range, shallow depth, and relatively sheltered geographic location make it an ideal environment for growth. Like the garden plants more familiar to many people, not all of them bloom at the same time. There are some, however, that thrive year-round or nearly so, such as kelp, rockweed, bladder wrack, and Irish moss. Others may appear for days, weeks, or months. About 250 species have been documented in Long Island Sound by diligent collectors over the years. Some are found in hard-to-access places–for example, inside a blade of eelgrass, or within a mollusk shell. The species included here are generally common ones that you may readily encounter, plus a few that have unusual or noteworthy features.

Biologists put the large seaweeds into three groups according to their dominant pigments–Chlorophyta, Phaeophyta, and Rhodophyta–or simply, green, brown, and red. All contain chlorophyll, and carry out photosynthesis, but the green color is masked in other species by the additional brown or red pigments. These pigments absorb various frequencies of light. The limited light available at various depths in coastal waters determines the depth, or zone, in which the algae can be found. In general, greens are closest to shore, and thus highest in elevation, browns are in the intertidal (=littoral) zone and subtidal zone, and reds both farthest down and farthest from shore. There is of course some overlap.

Most seaweeds attach to rocks or other hard surfaces by means of a structure called a holdfast, but some float freely or form mats. Seaweeds provide habitat, food, and shelter for a number of aquatic animals. In the process of photosynthesis, they produce oxygen as a byproduct, and thus help to aerate the waters.

The algae are structurally much more simple than land plants; they do not have true roots, stems, or leaves. They are thought to represent the evolutionary ancestors of all the terrestrial plants, however. Despite this simplicity, many seaweeds, particularly the reds, have complex and fascinating life histories and reproductive structures. This guide will not go into detail on that topic, but the bibiography at the end will assist those who would like to delve further into the subject.

The diversity of forms that seaweeds take is remarkable. Most people would not readily recognize some species as algae. For example, some form thin crusts, others look like flat sheets, hair, spots, threads, branching shrubs, rubbery blobs, or delicate feathers.

The incredible range of structure, color, and forms of the seaweeds is rivalled by the number of utilitarian uses that humans have found for them, or from material extracted from them.

Colloidal extracts from seaweeds are used in commercial food production as thickening or stabilizing agents. Carrageenan, for example, found in Irish moss (*Chondrus*), makes ice cream, toothpaste, and many other products smooth and creamy. Another marine colloid, alginate, comes from portions of kelp (*Laminaria*) and the rockweeds (*Fucus* and *Ascophyllum*). Alginate is used in syrups and fillings as well as for dental impressions and coatings for paper, film, medications, and fabric. A third colloid, agar, (in *Gracilaria* and *Gelidium*), is used in fruit and cheese fillings as well as medium in laboratories for culturing organisms and even gel for DNA "fingerprinting" using electrophoresis.

C.R. Janiak

Agar, from red algae, is placed in petri dishes to grow cultures in laboratories. Its nutrients make it an ideal medium.

Seaweeds can also be sea vegetables, and are a wonderful source of vitamins, protein, and trace nutrients. Kelp, the ruffly long brown alga that children love to drape themselves in at the beach, is the star of the show; the largest seaweed of the Sound. Once collected and dried, kelp is used as soup stock or sea vegetable by many peoples of the world. In New England, kelp has traditionally been composted or liquified for use as a combination fertilizer and bug repellent in many a garden. While the giant kelp of the West Coast gets more popular acclaim, our local kelp, *Laminaria*, can grow as much as an inch-and-a-half per day in the winter!

Most people have seen or eaten nori (*Porphyra*), a red seaweed, in the form of sushi wrappers. Few, however, are aware that nori thrives wild in Long Island Sound. Nori is highly nutritious, with more vitamin C per unit than orange juice, more calcium than milk, and more protein than soybeans. Efforts to farm nori in the United States as an aquaculture industry have begun, but are still small, despite a recent evaluation of the world market at about $1.6 billion per year.

C. Yarish

Sushi is only one of the many products made from seaweed.

Seaweeds are also a source for producing many cosmetic and medicinal products. The same colloids that make food smooth can do the same for human skin, and can make coated medications easier to swallow. Natural antibiotics found in some algae, combined with the ability to absorb fluids, are useful in products such as surgical bandages.

One use of seaweed that has real potential for Long Island Sound and other valuable but threatened estuaries worldwide, is the use of seaweed to cleanse polluted waters. Connecticut Sea Grant-sponsored research by Dr. Charles Yarish and colleagues shows that seaweed acts as a very effective nutrient scrubber, taking up nitrogen and phosphorus, the very nutrients used in gardens for fertilizer, and converting them into healthy products. Aquaculture industries are learning the value of integrated aquaculture, that is, growing algae such as nori right alongside fish or shellfish. That way waters are cleaner and healthier, and yield more potential crops to harvest.

How to Use this Guide

This guide is meant for the curious wader or beach stroller, so scientific terms have largely been omitted, in favor of colorful photos. In most cases live specimens were collected, plopped into a dish or onto a piece of paper or a rock, and photographed. In cases where a fresh specimen was not readily available, we have used herbarium sheets generously provided by both the University of Connecticut at Stamford) and Connecticut College. The seaweeds are grouped according to their pigments, as green, brown, or red, as is traditional. Within those categories they are arranged alphabetically.

Each species has a page with "What it Looks Like", "Where to Find It", "When to Look" and "Notes". Names change fairly frequently as scientists discover more about the genetic origins of species, so if you don't find one you are looking for, check the index but also flip through the pages. It may appear under a newer name. Synonyms or earlier names, as well as closely related species not included, are provided under "Notes" on each page. The Notes section is also used to include interesting facts about some species.

Measurements are abbreviated as in. (inches), ft. (feet), cm. (centimeters), and m. (meters).

Probably the most fun way to use the guidebook is to have your own "scavenger hunt," checking off the species that you find. If you are curious about a particular seaweed that you have found, flip through the pages until you spot a picture and description that fits. There is tremendous variation in shape, color, and sizes of the algae, influenced by the prevailing environmental conditions. Or, if you'd like to play detective by following a set of observations and clues, use the simplified key in the back.

Sections on how to collect and preserve your specimens are included. Much can be learned about smaller species by using a hand lens for magnification. Some species can only be definitively identified by using a microscope, however. In some cases insets showing what structures might be observed with a microscope are included for the reader's enjoyment and education.

The diagram on the next page will show the reader what is meant by the habitat descriptions, such as spray zone, mid-intertidal, subtidal, etc.

For those who decide to learn more about the seaweeds, some classic textbooks are listed in the reference section. Included also are two widely used comprehensive taxonomic keys for the serious biologist. For anyone who has hopes for becoming an accomplished seaweed chef, or wants to cook up a dish for a class, an excellent inexpensive cookbook is also listed in the reference section.

Finally, photographers have been credited next to their photos, in case a reader wants to contact the author about the potential use of an image.

How to Collect Seaweed

Collecting seaweed is obviously best done at low tide, in wading boots, when more surfaces are exposed and accessible. Since Long Island Sound has two low (and two high) tides per day, you'll have many opportunities. Since most seaweeds attach to rocks, rocky coasts and other hard surfaces such as dock pilings are best; however, some species inhabit tidepools and the edges of marshes too. Beaches can be great places too. After a storm, great masses of algae may wash ashore. These masses, known as "drift" or "wrack" are the easy way to get many species without snorkeling or diving, but you'll have to do some work to carefully separate items you want to keep from the rest of entangled mass, and you may find fragments rather than whole organisms. Most public beaches, unfortunately, quickly remove algae that washes ashore. Thus, the less frequented areas may be more fruitful.

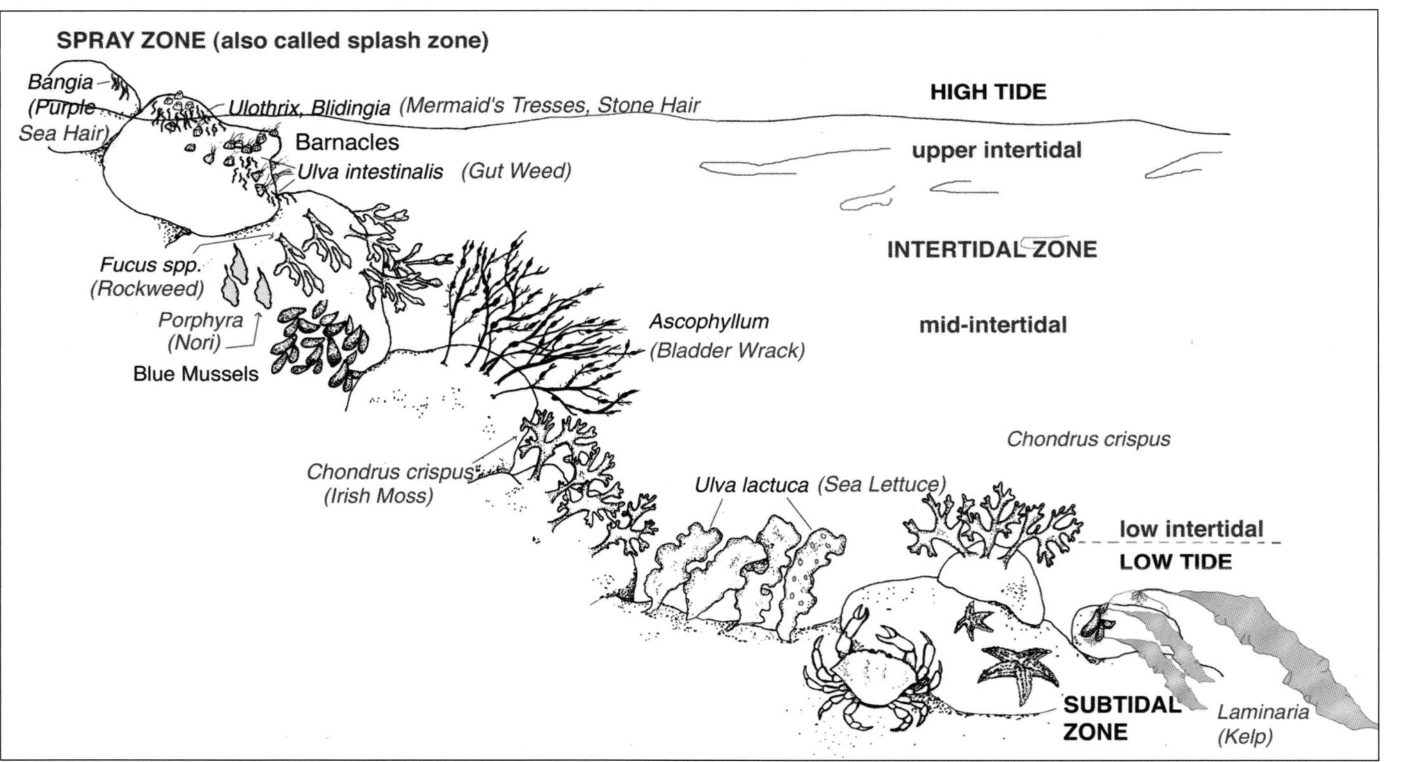

Zones occupied by the algae on a rocky shore, with a few representative species..

The first safety rule for collecting is, wear sturdy shoes or wading boots with good tread. If you're climbing on rocks, try to select dry ones as much as possible. The black or grey color that appears on the surface of many rocks is actually a coating of blue-green algae that can make them very slippery. The next rule is try to go with a buddy, in case you do slip and fall. You can walk on sandy or muddy bottom that is interspersed with rocks, but please try to avoid stepping on small animals such as grazing snails and young crabs.

If you are simply having fun getting to know the seaweeds, just take a notebook and start a checklist. Take photos too if you like but best to do that on shore, lest you slip and give your camera a dunking. However, if you want a permanent collection to refer to, however, or if you want to use a microscope, you'll need to take the specimens along with you. When you select a specimen, make note of where it was collected, the date, and any other relevant information such as habitat, wave conditions (exposed vs. sheltered or calm, etc.) This will go into your permanent record, and may prove valuable to future collectors. Carefully remove the specimen. If firmly attached to rock, you may need to use a utility knife or similar tool in order to detach the holdfast. Rinse in the seawater to remove any small animals or particles. It can be placed in a plastic bag with enough water to keep it moist, or wrapped in dampened paper towels. A collection bucket with ice in it is even better. After you leave the collection site, unless you can immediately process your seaweed, put all specimens in a cool place such as a refrigerator. They can stay for a couple days but will begin to smell as they deteriorate. Most can be frozen if absolutely necessary, but the structure after thawing will not be ideal and they will disintegrate soon afterwards.

How to Preserve Your Collection

To make a permanent collection, you will need to dry and press your algae onto acid-free paper. Herbarium sheets, which can be purchased from biological and archival suppliers, are best but good artist's paper or even index cards can be used too.

On the bottom right-hand corner of your paper, put the following information: first, the species you believe you have collected. If you don't know yet, you can add it later. On the next line(s), put the date and location where you collected the sample. Next, add any important additional information about the habitat, the condition of the water, etc. Finally, put your name, as the collector. You may also want to number your sheets for easy identification later on. This way of labelling the collection is a convention that botanists use. It makes it easy if someone wants to compare samples or ask questions later on. Since these collection sheets can potentially last for hundreds of years, it may be much later.

Put your sheet of paper into a large tray, or even in a tub. After it has been gently rinsed, arrange your specimen on the sheet so that it is attractive, and somewhat centered. The holdfast should be at the base, if there is one, as in nature. Now either immerse your sheet and specimen in a very thin covering of water, or gently spray or squirt water on it to spread out the branches. This is particularly important for the red algae, which may be a dense clump until you spread the branches. If you have a particularly bulky specimen, you may trim away some of the branches so that the basic structure is clearly seen.

Once your specimen looks the way you want it to, gently slide it out of the water and let it "drip dry" for a few minutes. While it is draining, start a "sandwich" by laying down a sheet of corrugated cardboard at least as large as your herbarium sheet, then add a sheet or two of blotting paper (or several layers of paper towels). Now put down your herbarium sheet with the specimen on top. If you have a piece of nylon net fabric, or scraps of old pantyhose handy, place it on top of the specimen. This step is optional but allows air to circulate nicely and keeps the specimen from molding or sticking to the next layer. Some professionals prefer sheets of waxed paper above the specimen. Finish your "sandwich" by reversing the order of the layers–blotting paper or paper towels, then another sheet of cardboard. Make as many "sandwiches" as you need to accommodate all of the specimens you wish to dry, and stack them. Once all of your "sandwiches" are assembled, you can put them in a press. The quick and dirty method is to place them in a cool environment and put a couple of heavy weights such as rocks on top. Or use a couple pieces of thin plywood or pegboard, strapped together with belts or bungy cords. The press in the photo can be purchased or replicated using pieces of scrap lumber and straps. Holes drilled in the top and bottom encourage air circulation, preventing mold. Handles can be added for convenience.

You may be wondering about the fact that no glue has been mentioned. The colloid substances that keep algae moist, and are so important to industries, act as a natural glue, oozing forth as the piece dries and adhering it to the paper. If by chance a specimen comes loose after drying, you can always touch it up with a clear-drying glue. Place in a well-ventilated area to dry.

In a couple of days, your specimens will be dry and ready to archive.

P. Van Patten

A press can be homemade or purchased. This photo shows a popular traditional design.

Keep them in a place where the temperature stays pretty constant and doesn't overheat. One of the marvelous things about seaweeds is that, if you ever need to examine a structure microscopically, you can always cut a small piece from your dried specimen later on, reconstitute it with water, and observe the structures nearly "as good as new".

P. Van Patten

Luckily, seaweed collecting is not limited to students–it can be a satisfying, fun, and an educational hobby for men, women, and children of all ages–anyone who loves to be outdoors on the shore. Long Island Sound has approximately 250 species, due to its variety of diverse habitats. Shown: Becky Gladych, collecting.

Seaweeds have various pigments that use different wavelengths of light to carry on photosynthesis. Species also vary in their requirements for temperature, salinity, and moisture. As a result, seaweeds tend to occur in fairly distinct horizontal belts or zones, according to which micro-habitat best meets the requirements of the species. Green algae use chlorophyll as the dominant pigment, so tend to be highest in elevation, as here on a large rock with barnacles. Brown algae such as *Fucus* have accessory brown and yellow pigments which can use green light, and are further down in the intertidal zone but also include *Laminaria*, which is always submerged. Red algae have red and blue pigments that use primarily blue light, the last to be extinguished as water depth increases, and so are found in the deepest locations that still receive light from above. There are many overlaps and exceptions. *Porphyra*, a red seaweed, tends to occur high in the intertidal zone, compared to most other reds, and can extend into the subtidal zone. In this photo we see *Blidingia* above and in the barnacle belt, *Porphyra* below it and at the bottom of the barnacles, and *Fucus spiralis* in the lowermost belt, a typical Long Island Sound upper to mid-intertidal situation.

C. Yarish

Seaweed Online

The University of Connecticut Jeremy Richard Library maintains the Benthic Marine Algal Herbarium of Long Island Sound online. There, pressed specimens from various collectors and locations have been scanned as digital images and are available to Internet users. You can compare your identification to the images of specimens others have collected, and learn more about the species. Access the web site at http://www.algae.uconn.edu. Another excellent resource for checking your identifications is AlgaeBase, maintained by the National University of Ireland, at http://www.algaebase.org.

An herbarium sheet displays your specimen, and information about it, with a label in lower right corner. Labels can be handwritten on the sheet, or prepared separately and glued on.

Image with ruler for scale from *http://www.algae.uconn.edu.* *Palmaria palmata* specimen collected from Long Island Sound by the author.

Green Seaweeds

Seaweeds classified in the green division (Chlorophyta) get their color from chlorophylls *a* and *b*, the green pigments. They may also contain secondary yellow-orange pigments. There are about 7,000 species of green algae in the world, of which about 900 are marine. Of course you won't find that many in Long Island Sound alone, but there are approximately 60-75 separate species. The numbers change as species come and go.

Most of the green seaweeds are found from the spray zone (the uppermost part of the shore that the water ever reaches) to the mid-intertidal (the part that is submerged when the tide is high and exposed when it is low.)

The color, which may range from pale yellow, "spring" green to bright "grass" green to dark, "forest" green, is your best indication that you have a green seaweed. However, to tell for sure whether or not your specimen is classified as green, you can try staining a portion with a drop of iodine solution. The starch products from photosynthesis that are stored in the cells will turn dark purple to black. If nothing happens, then your specimen is likely classified as brown. Most brown seaweeds that appear green have an olive coloration, but some brown and reds turn green as they deteriorate after washing ashore, particularly on a sunny day. Some red species also have a rare green form too, as a result of genetic mutation. *Gracilaria* is one example.

C.R. Janiak

P. Van Patten

Acrosiphonia arcta

Green Rope, Green Pompoms

What it Looks Like

Filaments form round dark green tufts; with diameters up to nearly 6 inches or more, but usually smaller, as seen above. Young ones (inset) are soft; older specimens may be somewhat stiff and may have spines or hook-shaped branches. The filaments in mature tufts may be twisted like rope.

Where to Find It

Open, exposed shore; low intertidal zone and tide pools.

When to Look

Throughout the Sound in early spring, but may appear in late winter north of LIS.

Notes

This alga alarmed beachgoers when many washed up en masse onto Ocean Beach, New London, Connecticut, in May, 2005. They were present there, and at other beaches on Long Island Sound, for only for a few days. One observer said "it looked like the beach had green chicken pox." Like the vast majority of seaweeds, they are completely harmless and may be ecologically beneficial.

P. Van Patten

Blidingia minima
Stone Hair

What it Looks Like

Very small, tubular, bright yellowish-green filaments that grow in large carpet-like colonies. May grow to a maximum 2-8 in. (5-20 cm.) long. The width is similar to human hair, thus the common name.

Where to Find It

Upper intertidal zone, on rocks or driftwood.

When to Look

Year-round.

Notes

These were photographed at Avery Point, in Groton, Connecticut but can be found all over the upper intertidal areas around Long Island Sound. They are slippery so be careful walking on rocks! Snails eat it, for its carbohydrates.

C.R. Janiak

J. Sears

Bryopsis plumosa
Green Sea Fern, Sea Moss

What it Looks Like
Delicate, feathery, branches in a regular pattern, may grow individually or in clumps. Resembles a fern. May grow to about 4 in. (10 cm.). Color is light to dark green but base may appear brown. May have a triangular or "Christmas tree" shape. Branching tends to be flattened.

Where to Find It
Protected areas, on rocks or shells in the subtidal zone, or on larger algae.

When to Look
Spring to fall best, but can be year-round.

P. Van Patten

Notes
Similar, related species also in LIS: *Bryopsis hypnoides*. Branching is irregular; it is larger and may be a darker shade of green.

P. Van Patten

Linear cells magnified 40x

Chaetomorpha linum
Green Thread, Green Brillo®

What it Looks Like

Resembles a tangled ball or clump of green thread. Individual strands are thicker than human hair, and about 5 times, or more, as long, with many curls. Size of clump varies.

Where to Find It

Upper intertidal zone, on rocks or driftwood, often entangled with other algae.

C..R. Janiak

When to Look

Year-round.

Notes

Novices spotting this may think someone has been polluting the waters by discarding their sewing debris. The cells are in a single, linear arrangement as the inset shows; the mass is made of single filaments rather than densely branching ones– like a permanent wave hairdo that went frizzy. A related common species: *C. melagonium* is larger, with huge cells visible to the naked eye.

P. Van Patten

Cladophora cells, magnified.

Cladophora sericea
Green Tuft, Green Hair

What it Looks Like

Soft yellowish-green- o tan, or bright green tufts; may also be lime-green or blue-green. Filamentous branches and branchlets, pretty straight. Looks fuzzy or cloudy when viewed in the water. Grows to about 10 in. (25 cm).

Where to Find It

Open coast, estuaries, somewhat exposed areas, intertidal and subtidal down to nearly 10 ft. (3 m.) and in tide pools; usually attached to other algae or rock but may form free mats.

When to Look

Year-round

Notes

C. Yarish

Related species: *C. albida* is slightly smaller, very common, and has curved branches. There are also a number of freshwater and brackish species of *Cladophora*, which many boaters consider a nuisance; extensive mats form during a bloom condition (above). *Cladophora* balls are popular as an aquarium food product in Asia.

M. Van Patten

C.R. Janiak

A cross-section under the microscope reveals the "utricles", structures filled with cytoplasm that make *Codium* spongy.

Codium fragile

Dead-man's Fingers, Oyster Thief, Green Fleece, "Sputnik weed", Staghorn Weed

What it Looks Like

Branches. about pencil-thick, fork into Y-shapes; looks and feels spongy, like wet felt; gets dense and bushy as it grows yet is buoyant. Older ones are dark green; young growth is lighter and more yellow-green, as in the larger top photo. May grow up to 3 ft. (91 cm.). The holdfast is an irregular shape.

Where to Find It

Upper intertidal zone, on rocks or driftwood. Grows well in either nutrient-rich or nutrient-poor environments. See habitat photo (right).

C. Yarish

When to Look

Year-round.

Notes

This species, an invader from Europe, was spotted near Orient Point, Long Island NY in 1957, at the same time the Soviets launched their Sputnik satellite. The buoyancy comes from oxygen, produced during photosynthesis, that cannot escape the spongy body. It is despised by shellfishermen because it can settle on oyster shells and can lift them out of their beds. It's edible but mushy.

Thickness comparison

Monostroma grevillei

Sea Cellophane

What it Looks Like

Flat blade; very thin (one-cell thickness) like tissue paper. It is half the thickness of sea lettuce (*Ulva*). Color may be yellow-green to light olive. Sac-like when young, opening to flat, often wedge or fan-shaped, sheets. Grows to about 6 in. (15 cm.) long. During summer, it forms a tiny single-cell stage that appears as a green stain or thin crust on rocks or shells.

Where to Find It

Open coast, estuaries and bays, low intertidal zone.

When to Look

Most common from late winter to spring. Prefers cool water.

Notes

See inset for an idea of the thickness of *Monostroma* compared to a fragment of *Ulva lactuca*. If you can see your fingerprint through the blade, it is probably *Monostroma*. Some related species and recent name changes due to reclassification: *M. oxyspermum (=Gayralia oxysperma, Ulvaria oxysperma), M. pulchrum (=Protomonostroma undulatum), M. leptodermum (=Kornmannia leptoderma).*

C. Yarish

P. Van Patten

Prasiola stipitata
Short Sea Lettuce

What it Looks Like

Minute, less than 1 cm. (about 3/8") tall and wide. Flat blade but edges may be curled over, on one side or both, like a mouse ear. Color is bright to dark green.

Where to Find It

In the spray zone (above average high tide). Tends to grow on rocks that sea birds frequent, where the bird droppings are. Often on one side of rock only. Also in high tide pools.

When to Look

Year-round.

Notes

Not easy to find, but isn't the hunt half the fun? The name comes from Latin words that mean "Leek-green".

P. Van Patten

Rhizoclonium riparium

Cells viewed under microscope

What it Looks Like

Very small, soft, entangled hairlike filaments, with single cells attached end-to-end in a line, sparingly branched. Color varies from yellowish green to medium green; may form mats. With a hand lens, or better yet a microscope, you may be able to see root-like projections, rhizoids, that give this species its Latin name.

Where to Find It

Open coast, estuaries; intertidal zone or below, attached to rocks or in mats in tide pools and marshy areas.

When to Look

Year-round.

Notes

The sample photographed was a mat of entangled strands found in the intertidal zone where the beach was covered with dune grass and small tide pools had formed. A similar species, *Rhizoclonium tortuosum,* is darker green or bluish-green and slightly thicker.

P. Van Patten

C. Yarish

Ulva intestinalis habitat

Ulva intestinalis

Gut Weed, Green String Lettuce

What it Looks Like

Bright "spring" yellow-green to bright green color. Unbranched tube with undulating edges, in LIS it may grow up to 16 in. or so (40 cm.) long by about a half inch (1-2 cm.) wide at most. Width of the tube may vary along its length with some twisting, resembling animal intestines in shape. The tube is inflated with gas, helping it to be erect under water.

Where to Find It

Intertidal and subtidal; on rocks or in tide pools; single or in groups.

When to Look

Year-round.

Notes

This species was better known historically as *Enteromorpha*, but recent determination showed that most green blade algae are genetically the same as *Ulva*, sea lettuce. Related species: *U. linza; U. prolifera; U. flexuosa, U. compressa, etc.,* all formerly *Enteromorpha*. The name *Enteromorpha intestinalis* is still widely used in many texts.

Ulva lactuca

Sea Lettuce

Cells, 2 layers thick, viewed under microscope, magnified 400x.

What it Looks Like

Bright green to "grass green". Single flat blade, firm (2 cells thick). May have round or oval holes, or not. May grow up to nearly 24 in. (60 cm.), although smaller specimens (~12 in.) are more common.

Where to Find It

Open coast, estuaries and bays; intertidal zone down to about 36 in. (91 cm.) deep. Thrives where nutrients are abundant, so may be found in polluted waters or near sewage outfalls.

When to Look

Year-round; common.

Notes

Can be rinsed and chopped for inclusion in salads as a green vegetable. Also used as nutritious filler in sausage meat, and food for abalone in land-based integrated aquaculture systems. Provides food for a number of small animals. *U. lactuca* is called an "indicator species" for environmental health. Too much of it in one place can indicate excess nitrogen in the water. Compare with *Monostroma*.

P. Van Patten

P. VanPatten

Ulva linza
Mini Sea Lettuce

What it Looks Like

Bright "spring" yellow-green color. Unbranched flattened tube with ruffly edges, up to more than 15 in. (40 cm.) long by 2 in. (5 cm.) wide. Under some circumstances, it may not be a tube, but rather flattened as though the tube unrolled.

Where to Find It

Intertidal and subtidal; on rocks; single or in groups.

When to Look

Year-round.

Notes

This species is quite common. You may have eaten this as garnish on "seaweed salad" in fancy restaurants. As in *Ulva intestinalis,* this species was formerly *Enteromorpha linza*, but now is recognized as *Ulva*. Related species: *U. intestinalis, U. compressa, U. prolifera, U. flexuosa.*

UConn Jeremy Richard Libraries

C. Yarish

Cells viewed under a compound microscope

Ulothrix flacca
Mermaid Tresses, Wooly Hair

What it Looks Like

Small, thin filaments, less than 4 in. (10 cm.) tall, unbranched. Color is bright to dark green; usually a bit darker than *Blidingia.* Branches are cylindrical but thinner than most human hair. Holdfast a basal cell; tufts may twist together into skeins.

Where to Find It

Open coast, estuaries, upper intertidal zone; attached to rocks or shells, or growing on larger algae such as *Fucus*.

When to Look

Year-round but common in winter and spring

C. Yarish

Habitat of *U. flacca*

Notes

Those lucky enough to have access to a microscope will see that the cells have chlorophyll-containing structures that are often shaped like an open bracelet (see inset).

Brown Seaweeds

There are approximately 1,500 known brown seaweeds in the world, of which the overwheming number, about 99%, are marine. In Long Island Sound, about 50-65 species have been observed. Most prefer cold waters. The name Phaeophyta comes from Greek words for "Brown Plant". These algae get their brown or tan color from pigments known as fucoxanthins. Of course, they also contain chlorophylls, and some of them actually appear more olive-green than brown. Some red algae may look quite brown, too, so if you don't find your specimen in this section, take a look at the reds, and vice-versa.

Like the green and red algae, brown seaweeds come in a great variety of sizes and forms. Forms include sheets, filaments, crusts, and branching shapes. Sizes vary from about 1/4 inch to 20 or even 30 feet or more, in the case of kelp. Kelp are both the largest and most structurally complex.

Some, such as the kelp and rockweed, are economically valuable in producing food and cosmetic products. They are not currently harvested in Long Island Sound, but are in many other parts of the world. Most provide important food and habitat for small grazing animals.

C. Yarish

Kelp is by far the largest and most complex seaweed in Long Island Sound. Barry Egan, left, and Dr. Charles Yarish, right, hold the brown kelp, *Laminaria longicruris* at Avery Point in Groton, Connecticut in 1988. *This species was recently re-named Saccharina longicruris.*

P. Van Patten, N. Balcom

Ascophyllum nodosum

Bladder Wrack, Sea Whistle, Knotted Wrack, Rockweed

C.R Janiak

What it Looks Like

Very large, coarse alga that drapes down from rocks between the low and high tide lines. Olive-green color, sometimes ranging to yellow or to quite dark brown-green; main axis is flattened. Size: up to about 39 inches (1 m). Single, large, firm, oval float bladders are conspicuous along the branches. Tips of branches often forked.

Where to Find It

Intertidal zone, on rocks; same zone as *Fucus* but in the places sheltered from heavy wave action.

When to Look

Year-round

Notes

You may call this common species "Old Man of the Sea" as it may live to be anywhere from 20 to several hundred years old. Each node produced indicates a year, like tree rings. Inflated tips contain eggs and sperm, which shed each spring.

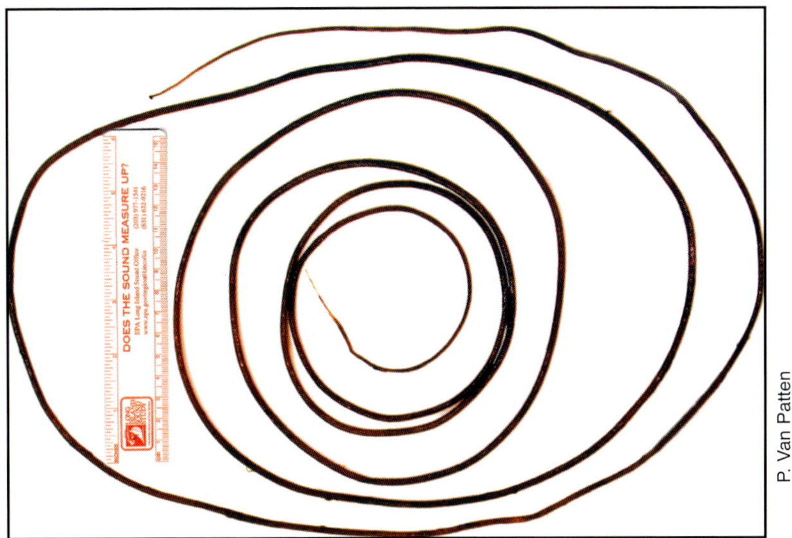

P. Van Patten

Chorda filum
Cord Weed, Shoestring Weed, Devil's Whip

What it Looks Like
Brown to dark brown, firm, smooth, cylindrical, and long; resembles a cord or thick string. It may grow to be up to 10 ft. (3 m.) long. The holdfast is a very inconspicuous disc. The cord becomes a hollow tube with age. Except for the tapered ends, the diameter remains pretty uniform, unlike *Scytosiphon*.

Where to Find It
Open coast, estuaries and bays, subtidal zone down to 34 ft. (10 m.) depth; often growing in clumps on pebbles.

When to Look
Spring to early fall.

Notes
A similar-looking species, now named *Halosiphon tomentosus,* formerly *Chorda tormentosa*, is characterized by a dense covering of small fuzzy hairs (see page 39).

P. Van Patten

Chordaria flagelliformis

Whipweed, Devil's Whip, Angel Hair, Brown Spaghetti

What it Looks Like

Very dark brown to black color; many cylindrical branches attached and curling out from a central cylindrical axis; the branches generally do not branch again. The diameter of the main branches may be similar to that of a pipe cleaner. May grow to about 1-2 ft. (30-60 cm.).

Where to Find It

Throughout Long Island Sound, open coast, estuaries; intertidal and subtidal zones. May be attached to rocks or other algae.

When to Look

Late spring and summer.

Notes

This alga was abundant on the eastern end and north shore of Long Island Sound in August, 2005.

UConn Jeremy Richard Library

Desmarestia viridis

Sour Weed, Stink Weed

What it Looks Like

Light golden brown to olive brown color; fading to dull green as it dries out. Many fine "hairs" on delicate opposite branches, although the hairs may vanish in the fall. Small disc holdfast. It dries greenish. It may grow to about 12 to 24 in. (30-60 cm.) tall.

Where to Find It

Open coast, estuaries; intertidal and subtidal zones and tide pools. May be attached to rocks or other algae or floating free. Sometimes forms dense meadows in areas with kelp.

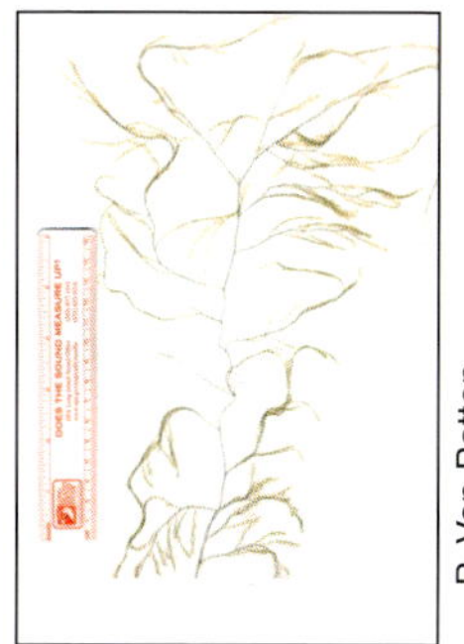

P. Van Patten

D. aculeata

When to Look

Most common in late winter to spring.

Notes

Beware! It may look harmless, but this alga exudes sulfuric acid that may damage the other specimens you have collected, if they are put in the same container. Related species, *D. aculeata* (right) has alternate branching.

P. Van Patten

C. Yarish

Magnified cells showing chloroplasts, the structures for photosynthesis.

Ectocarpus siliculosus

What it Looks Like

Ectocarpus is the smaller, tan-colored algae on *Fucus*, in the photo. Very soft, light brown to golden-brown tufts, many fine branches; tufts either floating free or attached. Looks cloudy when viewed underwater from above. Can be small as seen here or may grow to 20 in. (49 cm.).

Where to Find It

Intertidal and subtidal; usually growing on large brown algae such as *Fucus* or *Chordaria*.

C. Yarish

Ectocarpus siliculosus, free-floating, underwater.

When to Look

Year-round

Notes

Very common. It could be confused with *Cladophora* if not for the color. The reproductive structures, seen with a microscope, resemble tiny cat-tails. It's easy to confuse with *Pilayella* or *Hincksia* (see entries), or a related species not shown, *Ectocarpus fasciculatus*. *E. fasciculatus* is smaller and less common in the Sound.

P. Van Patten

C.R. Janiak

Elachistea fucicola
Troll's Hair, Pincushion Weed

What it Looks Like

Small, roundish, firm and erect tufts, like a little pincushion, growing on *Fucus* blades. Usually less than an inch (2.54 cm.) high. Branched only at the base, where you should see or feel a small "knob".

Where to Find It

Intertidal, usually growing on large brown algae such as the *Fucus* shown here.

When to Look

Year-round

Notes

Common, almost always on *Fucus*, but occasionally on *Ascophyllum or* other species. The species name means "*Fucus*-inhabiting". The inset is magnified a bit and shows the specimen on the midrib of a blade of a single blade of *Fucus.*

P. Van Patten

Fucus distichus

Rockweed

What it Looks Like

Large, mature specimens usually from a foot to nearly 3 ft (90 cm.) long; strap-like flattened branches with midrib (raised line down the center) like the more common rockweed on page 38, except there are no paired air bladders along the blades. The holdfast is small and disc-shaped. Receptacles at the branch tips are long and pointed, or blunt. Color is dark green-brown or olive-brown; under some conditions reddish brown.

Where to Find It

Open coast, intertidal and just below. Growing on rocks with other *Fucus* or *Ascophyllum nodosum.*

When to Look

Year-round.

Notes

There are two subspecies of *F. distichus, evanescens* and *edentatus. F. distichus ssp. edentatus* has very long pointy branch tips. *F. distichus ssp. evanescens* has shorter, blunt receptacles (tips), with a slight ridge. See similar species, *F. spiralis,* and *F. vesiculosus.*

C. Yarish

Fucus spiralis
Rockweed

What it Looks Like

A large, coarse, olive-green to brown colored alga that branches into two's. The blades are strap-like, have a midrib (or partial midrib) and twist. Length may be about 8 in. to 2 ft. (20-60 cm). The blades are about 1/2 to 3/4" (1-2 cm.) wide. This photo shows the bumpy spent reproductive tips, which have turned orange. It does not have paired air bladders (see p. 38).

Where to Find It

Intertidal rocks, generally at higher elevation than the other rockweeds and bladder wrack (*Ascophyllum*). May be with upper intertidal green seaweed such as *Ulva* or *Ulothrix.*

When to Look

Year-round.

Notes

Common throughout Long Island Sound. This species inhabits a higher elevation in the intertidal than *Fucus vesiculosus*, and does not grow as large.

C.R. Janiak

Fucus vesiculosus
Rockweed, Poppers

What it Looks Like

Olive-green to brown color, quite variable. Strap-like branches generally fork into two's., Has a midrib on flattened blades; paired float bladders filled with air along midrib (see inset). When reproductive, has rounded inflated receptacles at tips. The receptacles contain eggs and/or sperm, and become orange colored when ripe. A small, disc-shaped holdfast attaches it to rock. May typically grow from 1 to 3 ft. (30-91 cm) long; blades are about 1/3 to 3/4" wide.

Where to Find It

Open coast, intertidal zone and lower, draped on semi-exposed rocks. Tolerates low salinity levels at river mouths. Very common.

When to Look

Year-round; one of the most common species.

Notes

The large photo shows it in its habitat. Kids love to pop the air bladders along the blades–better than bubble wrap! Rockweed is traditionally used as a bed for shellfish in a New England clambake. It's made into rockweed tea, and used as fertilizer.

C.R. Janiak

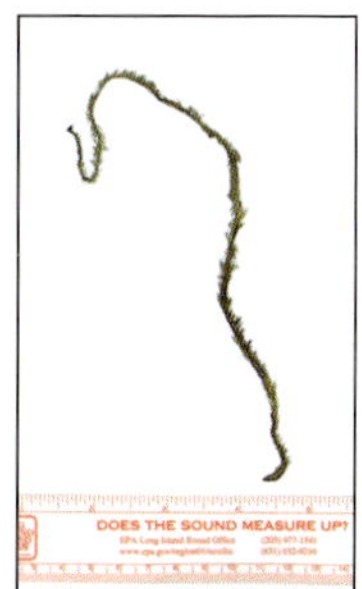

Halosiphon tomentosum

Hairy Shoelace

What it Looks Like

Single, unbranched cord, with many very short, soft, fuzzy hairs covering the main axis. Medium to dark brown but turns greenish when dried. May grow to be more than 3 feet (91 cm.) long. The big photo was taken with a zoom lens and shows parts of 3 individuals side by side; the inset shows a single specimen.

Where to Find It

Open coast, estuaries, cool places, attached to rocks, shells, and other algae.

When to Look

Late winter through spring.

Notes

Common throughout Long Island Sound in the winter. It is a cold-water loving species, also found northward, for example, in Newfoundland and Greenland. Formerly known as *Chorda tomentosa.* Compare to *Chorda filum* on page 31. The species name comes from a Latin adjective, *tomentose,* which means evenly covered with many short hairs.

UConn Jeremy Richard Library

Hincksia granulosa

What it Looks Like

Small, (up to 8 in., or 20 cm.) long, with filamentous, irregular, hair-like but slightly stiff branches. Diameters of main branches roughly size of typical human hair. Branchlets may recurve. May be in clusters or tufts. Color is golden brown to medium brown.

Where to Find It

Open coast, estuaries, attached to small stones, shells, and other algae.

When to Look

Late winter or early spring to early fall.

Notes

Found throughout Long Island Sound. Formerly known as *Giffordia granulosa.* This specimen is pressed. Closely related *Hincksia mitchelliae* is less stiff, more common, and is found year-round. You would need an expert with a microscope to study the reproductive organs to tell them apart for sure, but *mitchelliae* grows taller (to nearly 8 in., or 20 cm.) and is softer. They can also be difficult to distinguish from *Pilayella* or *Ectocarpus,* the other soft, golden brown, hairy, species.

www.carolscornwall.com

Laminaria digitata
Horsetail Kelp, Fingered Kelp, Kelp, Oarweed, Tangle

What it Looks Like

Very large, wide, thick brown blade divided into multiple strap-like sections from the U-shaped blade base. Has a claw-like holdfast, and a short stipe (see notes). It is usually not much larger than 39 in. (1 m) in Long Island Sound, but larger ones have been observed at North Dumpling Island. Young ones may not yet have fully divided and may appear more rounded.

Where to Find It

Subtidal, in wave-exposed areas. Not common here but has been found at Black Ledge, and the North Dumpling and Thimble Islands.

When to Look

In best condition in fall, winter, and spring.

Notes

The number of divided sections may vary. The stipe, or stem-like cylindrical part between the blade and the claw-like holdfast, can be marinated in vinegar and eaten as "sea pickles". The brown or golden brown color of kelp becomes green when it is decaying–and it will begin to smell, so you will want to keep your sample cool and press it quickly.

* Compare with other kelp species on pages 47-48.

P. Van Patten

Leathesia difformis
Sea Potato, Sea Cauliflower

What it Looks Like
Irregular shape, a rubbery-looking hollow mass; may be yellow to brown. May look like globs of yellow plastic, small lumpy brown potatoes, or something in between. The shape has also been likened to cauliflower, brains, and popcorn. Typically they are a few inches long or less.

Where to Find It
Open coast, intertidal zone. Nearly always growing on other algae, especially *Chondrus.*

C.R. Janiak

When to Look
Early spring to summer.

Notes
This species tends to be yellow to pale brown when young but gets darker brown with maturity. You might mistake washed-up older ones on the beach for "doggie doo" until you notice that it's rubbery and hollow. A good example of the fascinating range of forms that algae can take.

P. Van Patten

Petalonia fascia
Mini Kelp, False Kelp

What it Looks Like
Single blades, golden-tan to dark brown color, tapers at both ends. May grow up to 8 in. (20 cm.) long to about an inch-and-a-half (4 cm.) wide. Like thin paper when dried.

C. Yarish

Where to Find It
Open coast, estuaries, intertidal and subtidal, zone. May be attached to larger algae.

When to Look
Late fall to early summer.

Notes
This is a very common species in Long Island Sound. It is sometimes confused with kelp because they are both brown flat blades. However, *Petalonia* does not have a stipe or claw-like holdfast. It may also be confused with *Punctaria.* It has a very different alternate life stage, in summer, when it masquerades as a brown crust that dots mollusk shells.

UConn Jeremy Richard Library

Pilayella littoralis

What it Looks Like

Bushy, soft, hairy filaments, medium or golden brown color. May grow quite large–up to 39 in. (1 m.), although the size shown here is more common in the Sound. Looks cloudy underwater.

Where to Find It

Open coast, estuaries, intertidal zone. May be attached to larger algae, particularly *Ascophyllum* or *Fucus.*

When to Look

Spring to late summer.

Notes

This is a very common species in Long Island Sound. Usually found attached to larger brown algae but also may float free or in mats. In the western Sound, *Pilayella* is reportedly the most frequent species growing on *Fucus*, but in the eastern Sound the species attached to *Fucus* is more likely to be a *Ceramium* or *Polysiphonia fucoides.*

P. Van Patten

C. Yarish

Punctaria plantaginea
Ribbon Weed, Dotted Weed

What it Looks Like

Firm, single blades, tapered at ends. They are considered flat but may be curled like bacon strips on the griddle, as here. Golden brown to very dark brown color. May be individuals, or part of a group. The blades are 4-7 cells thick. May grow up to about 15 in. (40 cm.) long or more but usually more like 8-10 in. (20-25 cm.) in the Sound. The color may darken with age to a very dark brown. Mature ones may have reproductive structures that appear as dots on the surface.

Where to Find It

Open coast, lower intertidal to subtidal.

When to Look

Mostly found in the summer.

Notes

Similar species: *Punctaria latifolia* looks extremely similar but is thinner--only 2-4 cells thick. They are sometimes confused with *Petalonia*. Athough brown, not green, it has been compared to *Ulva*, because it sometimes forms large flat sheets in coastal ponds.

C. Yarish

Littorina snails have a scraping mouth and thus can feed on *Ralfsia*, the round brown spots in this beautiful underwater photo.

Ralfsia verrucosa
Tarspot

What it Looks Like

A yellowish-brown or dark brown to blackish crust, adhering tightly to rock. Grows in a circular form when young, but becomes irregular and patchy as individuals age and glom together. Diameter may be anywhere from 1/4 inch (0.6 cm.) to 4 in. (10 cm.). The crust is about 1-2 mm thick (about the thickness of the edge of a nickel coin).

Where to Find It

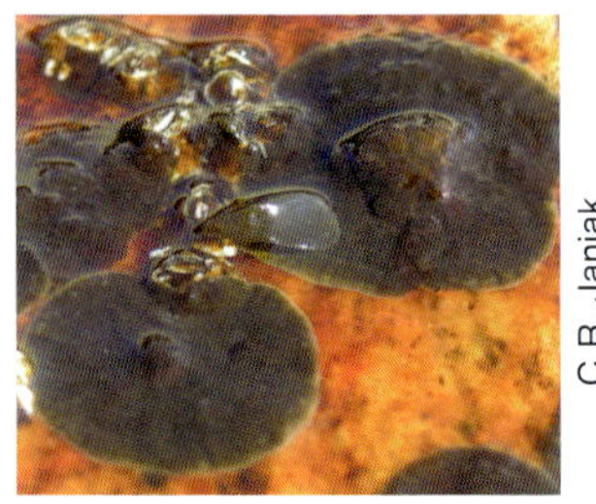

C R. Janiak

Open coast, estuaries; intertidal to subtidal zone.

When to Look

Mostly found in the summer

Notes

This species is common in New England. It really may look like a tar spot. If you want to collect it, you'll need a scraping tool like a utility knife–so don't slip! The *Littorina* snails shown feeding on *Ralfsia* in the photo are naturally equipped with a specialized scraping tongue on their foot, called a radula.

C. Yarish

P. Van Patten

Saccharina latissima, formerly *Laminaria saccharina*
Kelp, Sugar Kelp

What it Looks Like

Very large, flat but ruffly blade, much longer than wide, with cylindrical stipe and claw-like holdfast. If reproductive, may have a darker vertical bandin the center, like a stripe. It may or may not have rows of indentations on the blades. The blade texture and thickness has been described as "like a slice of ham from the deli".

Where to Find It

Open coast, subtidal, but washes up in quantity after storms.

When to Look

Year-round but fall, winter and spring are best. It tends to break up and disintegrate into fragments in late summer.

Notes

Kelp are the most complex and largest of the algae in Long Island Sound. Kids love to drape it around their necks at the beach, as a boa. It can be eaten as a vegetable used for won ton soup stock; harvested and composted for fertilizer too. Colloids extracted from it are used in many products such as syrup and fruit fillings. Kelp is rich in iodine, B-vitamins, and minerals. See also next page and *L. digitata.*

The stipe is hollow in cross-section.

A. Van Patten Kelly

Saccharina longicruris, formerly known as *Laminaria saccharina, ssp. longicruris*

Kelp, Atlantic Kelp

What it Looks Like

Very large, flat blade that is much longer than wide, attached to a usually hollow cylindrical stipe, ending in a claw-like holdfast, attached to rocks or mussel shells. May grow up to 36 feet (11 m) long. If reproductive, may have a darker central vertical band. Blade edges may or may not have ruffles.

Where to Find It

Open coast, subtidal, on rocks, or washed up after storms.

When to Look

Year-round but fall, winter and spring are best. It tends to break up and disintegrate into fragments in late summer.

Notes

S. longicruris, once considered the same species as L. latissima, is distinguished by its longer, hollow stipe. The long stipe can be cut into small pieces and marinated in vinegar to make "sea pickles". The stipe is also full of algin, which, as alginate, is used to make foods smooth and to coat fabric and paper. Kelp is rich in iodine, vitamins, and minerals. It can be eaten as part of a healthy diet.

NOAA Ocean Exploration

C.R. Janiak

Sargassum filipendula
Sargasso Weed, Gulf Weed

What it Looks Like

Large, light brown color, many branches that resemble long, narrow, serrated leaves. There are small round globes that look like gooseberries or small grapes–these are full of carbon dioxide, which makes it float.

Where to Find It

Enclosed embayments, subtidal. Not very common except in Niantic Bay, off Waterford, Connecticut, and the mouth of Mystic River, but has historically been reported infrequently in the western Sound.

When to Look

Year-round, summer best.

Notes

This species is not terribly common in the Sound but may be spreading. It is quite common in the Atlantic Ocean. It makes good habitat for seahorses, which sometimes come into Long Island Sound from the ocean via the warm waters of the Gulf Stream.

Scytosiphon lomentaria
Sausage Weed, Sea Sausage

What it Looks Like

Dark brown, tubular; unbranched filaments; with distinct contrictions form, making it resemble links of sausage. Younger ones may appear as a smooth tube. The tubes are filled with gas. During summer months, it may form a small brown crust, its alternate life form, seen as dark dots on mollusk shells and stones.

Where to Find It

Open coast, low intertidal to subtidal, and tide pools. May be in clumps, growing on other algae, or attached to rocks.

When to Look

Year-round.

Notes

The constrictions that give the sausage-link appearance become more pronounced with age. One thriving population in the western Sound was growing on an old tire. The alternate life stage is similar to that of *Petalonia.* Sometimes the combination of alternate forms of *Monostroma*, and *Scytosiphon* form an interesting pattern of green and brown on shells. Later on they become erect and grow taller, as shown.

Red Seaweeds

There are at least 5,000 species of red algae worldwide, most of which are marine, so they greatly outnumber the greens and browns. There are around 90 to 110 in Long Island Sound. As we've seen in the other divisions, the red algae come in many shapes, sizes, and forms. There are crusts, calcified coralline algae, filamentous forms, densely branching ones, and blades or sheets. The sizes range from the minute, barely visible species (see *Audouinella)* all the way up to 30 inches, for the recent invader, *Grateloupia.*

In general, though, red algae tend to be small in size, and often delicate and feathery. They have traditionally been pressed and incorporated as art into stationery, bookmarks, or framed pictures. Their red coloration, which ranges from light pink to deep red, purple, or black, comes from pigments known as phycoerythrins. These red pigments mask the chlorophyll. They are effective at absorbing and using green and blue light, which is available at greater depths in coastal waters, thus they generally occupy subtidal habitats.

Red algae are usually growing upon other, larger algae. Some are parasites. All have complex, fascinating life histories that are not detailed here, but make for interesting investigation for the devoted fan or biology student. Some resemble each other superficially and can only be truly identified to species level using a microscope. In this guide, representatives of the various forms are included, and some of the common species. In addition, a few that are newly invading Long Island Sound or have some particularly interesting feature are found.

Anyone who has enjoyed a trip to a sushi bar, or an ice-cream cone on a sunny day, has already appreciated the economically important red algae. *Porphyra*, better known as nori or laver, is the seaweed that, when pressed and dried, becomes nutritious sushi wrappers. Carrageenan and agar, two colloids that are used in a multitude of products including many food items, also come from red algae.

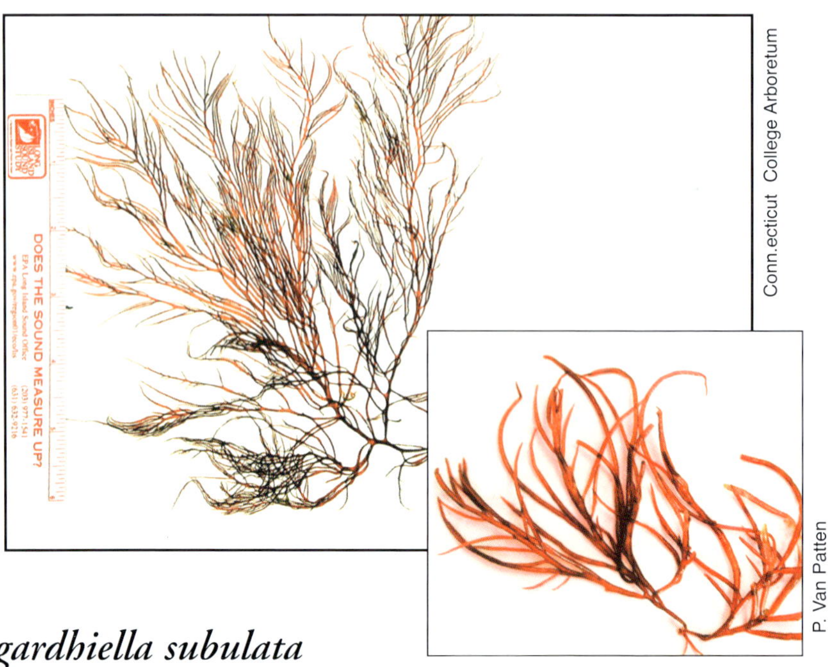

Conn.ecticut College Arboretum

P. Van Patten

Agardhiella subulata

Red Wooly Grass, Agardh's Red Weed

What it Looks Like

Deep rosy, red wine or maroon color. The branching habit is bushy, with tapering branches. It may grow to 12 in. (30 cm) tall; the branches are roughly the diameter of a pipe cleaner.

Where to Find It

Open coast, estuaries down to 30 ft. (10 m); often in drift.

C. Yarish

When to Look

Year-round.

Notes

This is one of the most common red algae in Long Island Sound. It is easil-confused with *Gracilaria. Agardhiella* contains antiviral compounds currently under study for medicinal use. *A. subulata* is sometimes habitat for young scallops, and part of the balanced diet of sand sharks and other animals. It is the same as *Agardhiella tenera*, a former name.

C.R. Janiak

Ahnfeltia plicata
Wire Weed

What it Looks Like

Fork-tipped branches, with diameters about the same or a bit thicker than paper clip wire. The branches may either be forked into two's or irregular, or a mix. It may be somewhat stiff, thus the common name, and entangled with other algae. Dark brown or purplish red to nearly black, sometimes with white or cream colored tips.

Where to Find It

Open coast, estuaries down to 30 ft. or more (10 m.); often in drift, or in rock crevices; sometimes in tide pools.

When to Look

Year-round, winter is a good time.

Notes

China and Russia cultivate this agar-containing perennial species as an agricultural industry to produce food and biopolymers. We do not know of any commercial harvest in the U.S.A. *Gelidium* and *Graciliaria* are more frequently harvested for agar than *Ahnfeltia*.

J. Foertch, J. Swenarton

40x

C.R. Janiak

Actual size

J. Foertch, J. Swenarton

100x

Antithamnion pectinatum
Red Sea Skein

What it Looks Like

Bright pink to pinky-red; very small, soft and delicate, like a tiny skein of embroidery floss; feathery branches with a very regular opposite branching pattern, with short tapered branchlets; crowded but flattened tips. Probably not much more than 1.25 in (3 cm) long. This small species is spectacularly beautiful under magnification because of its symmetry.

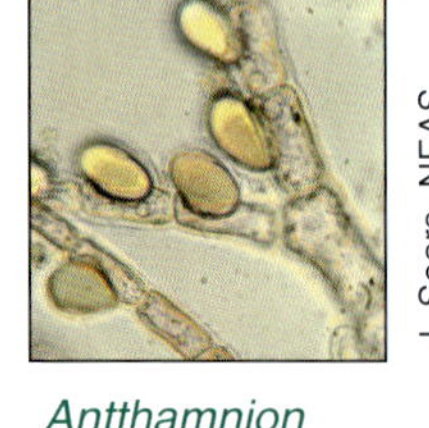

J. Sears, NEAS

Antthamnion gland cells, specialized structures for secretion or storage, can also help scientists confirm species identification.

Where to Find It

Open coast, subtidal; often epiphytic to (growing upon) other algae, such as *Chondrus* or *Corallina*.

When to Look

Year-round.

Notes

This species was first observed in the Sound at Millstone Point in 1986 by sharp-eyed expert Jim Foertch. A closely related lookalike species *Antithamnion cruciatum* (bottom inset), a native, is a rosy color and may grow a bit larger–up to 4 in. (10 cm.) long. *A. cruciatum* is often found on eelgrass.

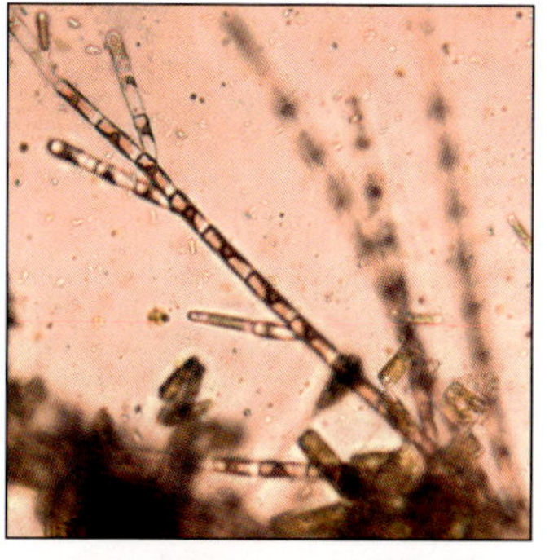

40x

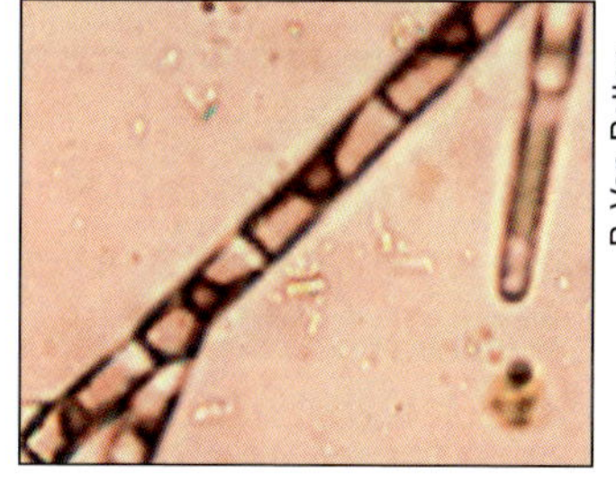

100x

P. Van Patten

Audouinella spp.

Distinctive linear cells.

What it Looks Like

This sparsely branched alga is quite tiny (within blue circles) and looks like minute pink to red threads or tufts, attached to larger algae, eelgrass, or hydroid animals. These were less than 1/4 in. (0.6 cm) in length. You may notice it growing on or mixed in with your other specimens, so it is included here to satisfy your curiosity. The tiny "threads", looking like lint on a sweater, were difficult to photograph in the actual size, but you can see the inset photos how lovely it is under the microscope.

Where to Find It

Intertidal to subtidal zone, attached to larger algae or eelgrass, often on the margins of the blades or on filaments.

When to Look

Year-round.

Notes

These were observed attached to *Polysiphonia,* in turn attached to *Ascophyllum.* There are several species that are difficult to distinguish, including some freshwater species, in addition to the estuarine and marine. Some of them form larger, round, pale pink tufts, looking much different from this tiny red "lint".

C. Yarish

Bangia atropurpureum
Black Sea Hair, Purple Sea Hair

What it Looks Like

Purplish red to black color, soft, slippery, draped on boulders. Single filaments. When abundant, it has been described as resembling dark hair on a balding head, or a wet Irish setter's coat. May be quite small, like "scum", or larger specimens up to to 6 in. (15 cm.) long.

Where to Find It

On rocks in the upper intertidal zone, right under the dark band of microscopic algae that marks the "spray zone". It is common throughout the Long Island Sound region and beyond.

When to Look

Spring and summer best.

Notes

Scientists who study evolutionary relationships believe that this slippery and persistent species was one of the earliest life forms to populate the Earth. Look how successful it has been!

UConn Jeremy Richard Library

P. Van Patten

Tightly coiled hooks appear along some branches. Magnified 40x.

Bonnemaisonia hamifera

Hooked Red Weed

What it Looks Like

Small, typically about 4 in., (10 cm.) tall, highly branched, delicate. May have a “Christmas tree” shape. Color is bright pinkish red to red. When examined very closely, one can see tiny distinct tightly coiled hooks along some of the branches.

Where to Find It

Open coasts,subtidal zone, attached to rocks or shells or growing on other algae. Often tangled up with other algae by means of the hooks.

When to Look

Year-round but prefers cold temperatures.

Notes

By far, the distinguishing characteristic of this species is the coiled hooks (inset), which help it attach to other algae. Unlike some other species with hooks, these are found along the branches, not only at the tips. Compare to *Hypnea* and *Cystoclonium*. This species also has an alternate life stage as a wooly pink to red-brown matted crust, called "*Trailliella intricata*". This mysterious crust, long thought to be a separate species, usually forms in the summer months. Compare to crust of *Scytosiphon*.

P. VanPatten

Callithamnion tetragonum

Beauty Weed

What it Looks Like

Dark red to brownish, highly branched, "spongy" yet delicate. It may be shaped like a pyramid. Densely branched and re-branched, with tapering tips. May grow to about 8 in. (20 cm.) long.

Where to Find It

Lower intertidal and subtidal zones, common throughout LIS. Often growing on other algae or eelgrass.

When to Look

Year-round.

Notes

"Tetragonum" means that its reproductive structures are divided into 4 parts, typical of red algae. A less common, closely related species, *Callithamnion corymbosum* (inset), tends to be smaller (2.5 in., 6 cm), brighter colored, and flat-topped, with very dense tips. Both are good choices for seaweed art, such as bookmarks or cards.

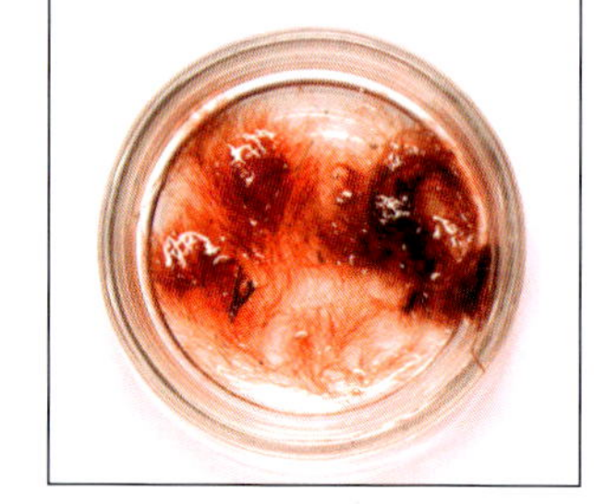

P. VanPatten, M. Lyons

This related species, *Callithamnion corymbosum*, is brightly colored, and is less common.

G. Saunders lab

Callophyllis cristata = *Euthora cristata*
Lacy Red Weed

What it Looks Like

Pink to rosy color; flat branches that divide repeatedly as they decrease in size–like a fractal pattern. May be bushy but usually the whole thing doesn't grow much larger than 2 in. (5 cm.) in the Sound.

Where to Find It

Subtidal zone, deep water, attached to rocks or base of kelp or as an understory below kelp canopy. (rare)

When to Look

In cool months, in deep, cool waters.

Notes

This cold-water perennial species has become quite rare in Long Island Sound in recent years. Perhaps its range has moved northward as the Sound has warmed. You may be lucky enough to find one, but it will be equivalent to finding a four-leaf clover. This lovely species was formerly known as *Euthora cristata,* and appears under that name in many keys.

Since it is rare, perhaps best to take a photo with an underwater camera and note the location and date, rather than collect for pressing.

Caloglossa leprieurii

Tetrasporangia (4-part reproductive structures) on the tips of *Caloglossa*, highly magnified.

What it Looks Like

Has a central rib; branch tips fork into two's; flat blades have regular constrictions that form small lance-shaped segments, about 1/4 inch (0.6 cm) long. Grows to 1.5 to 2.4 in. (4-5 cm.) wide. Color varies widely depending on depth of water, may appear dark green/gray, violet, or brownish red. Slippery.

Where to Find It

Sheltered, warm, brackish waters, near river mouths, on woodwork and pilings, and edges of salt marshes, from Connecticut River to western LIS.

Caloglossa (center) and other algae at the base of salt marsh grass.

When to Look

Year-round but summer best.

Notes

This warm water species is not generally seen in Eastern Long Island Sound. In the habitat photo, it is growing at the base of *Spartina alterniflora*, tall salt marsh grass. It tolerates very low salinity.

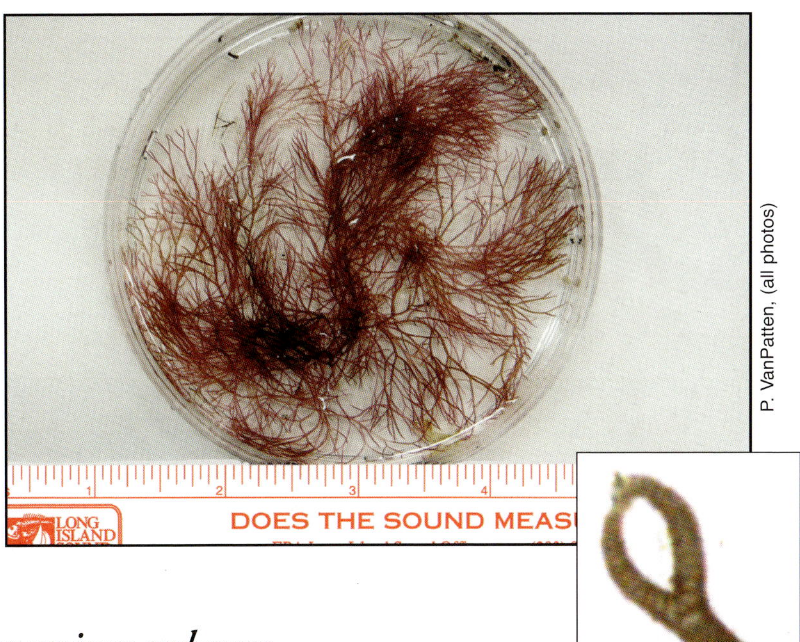

P. VanPatten, (all photos)

Crab claw-like pincer tips on *Ceramium rubrum* branches, magnified 20x.

Ceramium rubrum

Banded Weed

What it Looks Like

Dark red to red-brown, highly branched and somewhat coarse. May grow up to about 15 in. (40 cm.) but often much smaller. A hand lens will reveal pincer-type incurved tips (see inset). Shows distinct dark bands (actually clusters of many tiny cells) across branches when viewed with a magnifying lens.

Where to Find It

Lower intertidal to subtidal zone; common throughout LIS. Often attached to larger algae, particularly *Codium* (on *Fucus* here).

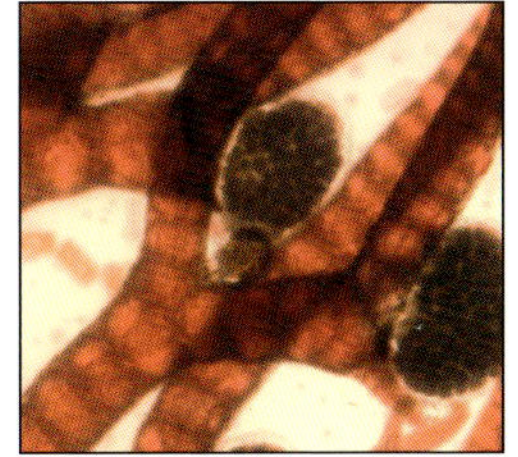

The oval, brown object in the center is the cystocarp, a female reproductive structure. Note distinct bands (40x).

When to Look

Year-round.

Notes

This is the most common *Ceramium* species in Long Island Sound, and is currently considered the same as *Ceramium virgatum*. Another common *Ceramium* is *C. diaphanum*, which is a dull red color and has forceps-like forked tips. Its bands are thicker than the branch, and so protrude, like collars.

P. VanPatten

A branch tip, magnified microscopically 80x.

Champia parvula
Barrel Weed

What it Looks Like

Red to brownish red tufts, abundantly branched. Tapering branches are cylindrical and alternate; tends to have a rounded overall shape. When you use a hand lens or just look very closely, you can see that the branches have rounded, short and plump segments that resemble little stacked tubs or a string of barrel beads, thus the common name. Branch tips are rounded. Size may range from just over an inch to about 3 in. (3-7 cm.) tall.

Where to Find It

Open coast, estuaries. May be floating free. Often in eelgrass beds.

When to Look

Late summer to fall is best.

Notes

The plump barrel-like segments, which make this such a charming species, are not individual cells. Rather, layers of tiny surface cells surround a much larger elongated cell, forming the visually darker bands between the "tubs".

This color results from drying in the sun. It's perfect for blanc mange pudding.

Chondrus crispus
Irish Moss

What it Looks Like

Shrub-like, bushy, densely branched and re-branched in forks from wedge-shaped base ending in short forked tips. Much variation in color and form. Color may be red to purple, brownish to black but bleaches to beige or ivory in the sun. In some tidal conditions the branches may be either thicker or narrower than in the top picture. When reproductive, has bumps on branch tips only. Grows up to about 3-6 in. (8-15 cm.).

Where to Find It

Open coast, estuaries. Subtidal zone, on submerged portions of rocks.

Chondrus crispus bed.

When to Look

Year-round.

Notes

Dried Irish moss is popular for making blanc mange, a traditional white pudding in New England, Canada, and Ireland. It is also harvested for its colloid, carrageenan, which is used commercially to thicken and smooth toothpaste and foods such as ice cream. Irish moss beds are good habitat for tautog (a fish).

P. VanPatten

C.R. H. Janiak

Reproductive bumps appear on tips only.

Coccotylus truncatus
Leaf Weed

What it Looks Like

A narrow main axis expands slightly in width as it branches out toward the red, wedge-shaped flat red blades, widest at the tips of the branches. Size range is from 4 - 8 in. (10 - 20 cm.) but smaller end of the scale is typical. If you use your imagination you may see red leaves on twigs.

Where to Find It

Open coast, estuary. Deeper waters of the subtidal zone, often the holdfasts of kelp

When to Look

Year-round, best in fall and winter.

Notes

This perennial species was formerly known as *Phyllophora truncata* or *P. membranifolia*. It is very similar in appearance to another alga, *Phyllophora pseudoceranoides*. One big difference is that this one has reproductive branchings only at the tips of the blades, whereas *P. pseudoceranoides* has the branchlings on other parts of the blades.

C.R. Janiak

S. Taylor & P. VanPatten

Corallina officinalis
Coral Weed

What it Looks Like

Very rigid, articulated exterior (like rock or bones); pink to rose or purplish-pink color but bleaches white when detached. Distinct opposite, angled, branching pattern (magnified inset). Cylindrical branches may have white tips. Grows to about 4 in. (10 cm.) long or a bit more.

Where to Find It

Open coast, common on submerged rock faces in lower intertidal into the subtidal zone; may be found with *Chondrus* or *Phymatolithon.*

When to Look

Year-round.

Notes

This tough but pretty, symmetrical alga gets its name from its resemblance to coral, due to the hard calcium-fortified exterior. The bone-like covering is calcium carbonate and/or magnesium carbonate. This can protect as armor against hungry grazing predators. The best way to preserve your specimen may be to keep it in an envelope glued to an herbarium sheet. A testiment to the durability of coralline algae is the fact that specimens collected by Charles Darwin during his famous voyages in the mid-19th century are still in great shape.

P. VanPatten

Grapevine-like coiled tendrils are the clue to identifying *C. purpureum*

Cystoclonium purpureum
Grapevine Weed

What it Looks Like

Large, light brownish red color; firm and bushy, quickly distinguished by the grapevine-like tendrils on the tips of some branches. Branches taper to a single cell at the tips. May grow up to 20 in. (50 cm.) tall.

Where to Find It

Open coast, mid-intertidal to subtidal

When to Look

Year-round, common

Notes

When reproductive, there are slight bumps or swellings along the main branches. They are not as pronounced as the distinct raised reproductive bumps on *Gracilaria*. Compare to *Bonnemaisonia* and *Hypnea*.

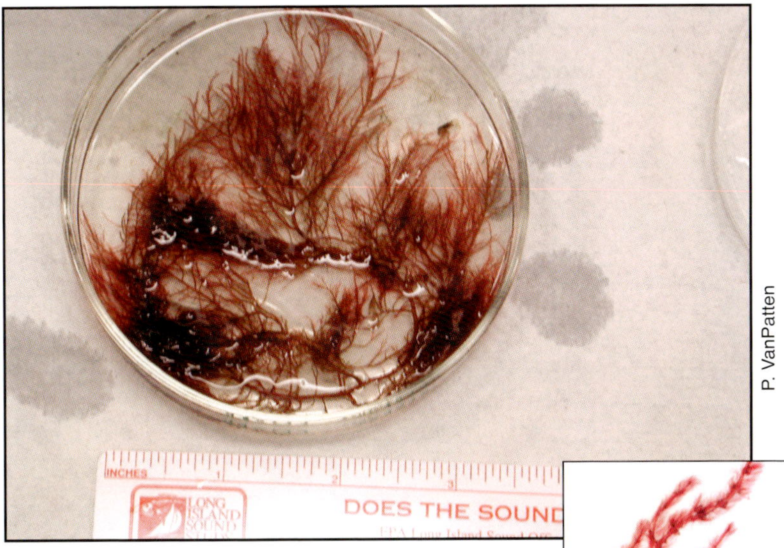

P. VanPatten

S. Taylor, Conn. College Arboretum

A pressed specimen.

Dasya baillouviana
Chenille Weed

What it Looks Like

Very feathery, delicately branched; soft fronds with many fine, short, soft "hairs" on larger branches. Shocking pink to bright red color. Described as resembling chenille pipe cleaners. May grow to nearly 30 in. (75 cm.) long.

Where to Find It

Open coast, subtidal; sometimes attached to other algae.

When to Look

Prefers warm warm waters, late spring to fall best.

Notes

A beauty that presses well on paper. A look through the microscope shows how the chenille effect is achieved–the very fine branchlets are in whorls around the larger branches.

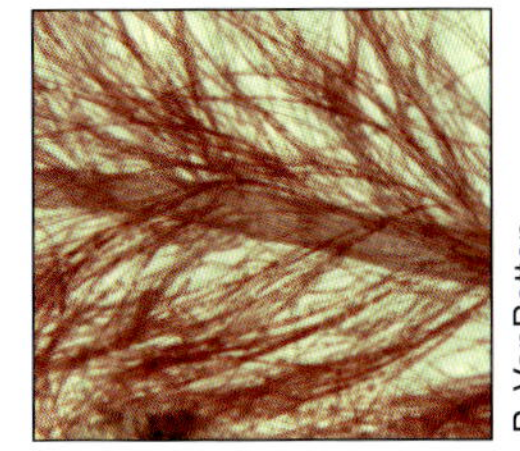

P. VanPatten

Whorled branchlets of *Dasya*.

C.R. Janiak

Dumontia contorta
Worm Weed

What it Looks Like

Red to red-brown, brick color; soft, unbranched or sparely branched, tubular filaments that twist. Base is a disc. They may be found in a cluster, as here. May grow to height of 4-20 in. (10-50 cm.). Branches may be typically close to the diameter of an M&M candy.

Where to Find It

Lower intertidal to subtidal, on rocks and shells; also in tide pools.

D.S. Janiak

When to Look

Late winter to early summer.

Notes

Named for a French botanist, Dumont, *Dumontia* is a cold-water species that is believed to have been introduced to Maine waters around 1913. It spread southward to Long Island Sound, where it seems to have reached its temperature limit. The *contorta* part of the name refers to the twisting of the branches. Compare with *Nemalion.*

UConn Jeremy Richard Library

Gelidium pusillum
Gelatin Weed, Gel Weed

What it Looks Like

Color is red as shown or more brownish-purple. Small stature, about 2 in. (5 cm.) tall, wiry, highly branched with many smaller branchlets. Branches are thin and typically end in three small trident-like branchlets with rounded tips. Feels slippery. Often in tufts.

Where to Find It

Mid-to low intertidal zone and edges of salt marshes.

When to Look

Year-round but likes warm water so summer is best. May shed tissue in cold weather or die back to its base to overwinter.

Notes

Gelidium species are grown in culture elsewhere as a source of commercial agar, as is *Gracilaria*. During World War II, *Gelidium* divers in California were reportedly exempt from the draft because agar was such an important national product. Agar is extracted by hot water, in which it dissolves, but in cold water it gels and firms. It is a white or yellowish color. It is sold in health food stores as powder or in a stick that feels like styrofoam, both of which dissolve in hot liquid. Great for thickening fruit fillings for baked goods.

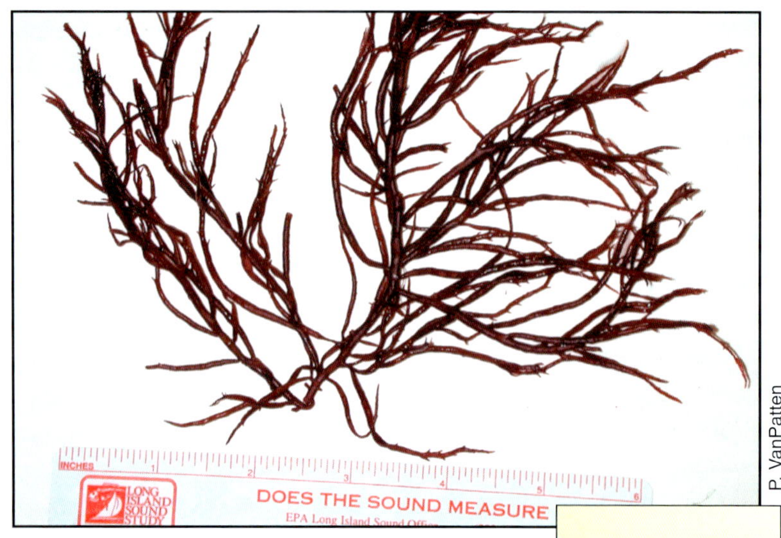

P. VanPatten

Gracilaria tikvahiae

Graceful Red Weed, Red Spaghetti

What it Looks Like

Dark red to purple, or wine color, but sometimes greenish, highly branched, with tapering projections; branches may have prominent raised round bumps along the branches if reproductive (see magnified inset).

Where to Find It

Open coast, subtidal, in warm waters.

When to Look

Summer.

Notes

In other parts of the world, this alga is economically important as a source of agar. It is farmed comercially in Hawaii and other places as an aquaculture product. Older, synonymous name: *Gracilaria verrucosa. See also Notes* under *Gelidium* for more on agar.

J. Preston

Gracilaria collected near a salt marsh in the eastern Sound.

P. Van Patten

Grateloupia turuturu

What it Looks (and feels) Like

Very large, flat, thick blade, with a deep red, burgundy or maroon color. It may grow to several feet or meters long. Blades have many irregular shapes. New individuals bud out vegetatively from the blade margins. Feels very slippery.

Where to Find It

Subtidal zone, competing with *Chondrus*. *Grateloupia* has been identified off Waterford and Groton, Connecticut, and Montauk, New York.

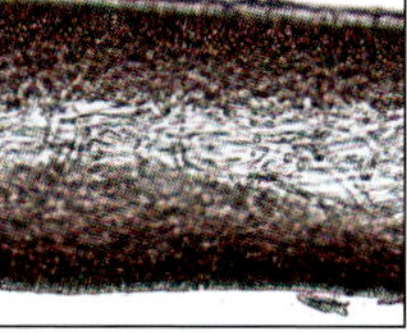

R. Gladych

A thin cross-section of a *Grateloupia* blade shows a layer of clear filaments in the center.

When to Look

Year-round.

Notes

This is a non-native invasive species from Japan that was identified in Long Island Sound in 2004, after reaching Narragansett Bay in 1997. It is expanding its range globally. *Grateloupia* is considered a nuisance, as it can displace Irish moss and possibly other species, from their native habitat. It is con-sidered useful in some places–cultured in the Philippines for its carrageenan, and also thought to have antiviral properties that deserve further study. It's easy to confuse with deep-water *Palmaria palmata,* which can be a similar size and color. *Grateloupia* is very slimy in comparison.

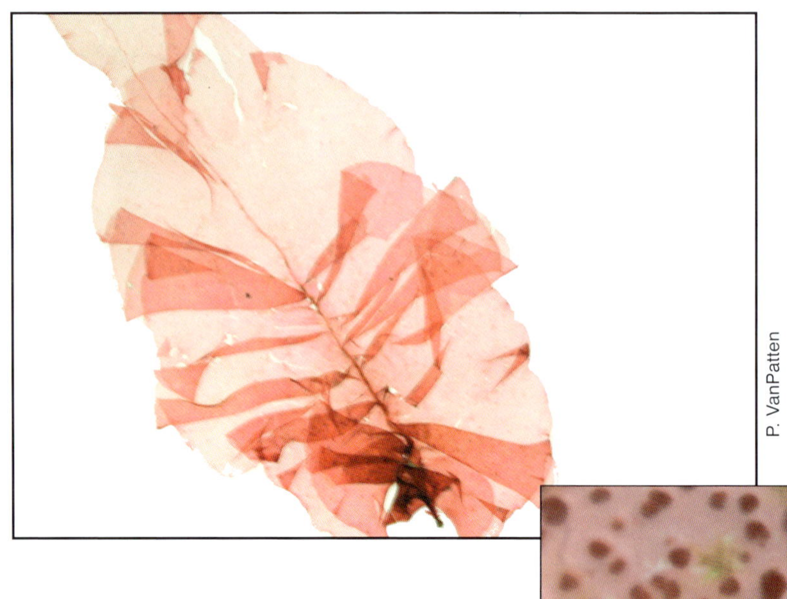

P. VanPatten

Reproductive "dots" from the surface, magnified 40x.

Grinnellia americana

Grinnell's Pink Leaf

What it Looks Like

A small, flat, very thin delicate blade with faint midrib; no branching. Color is pale to medium pink. May be an elongated oval shape or tapered, may roll or curl. May have tiny granular dots on the surface when reproductive. Becomes a "lifeguard-orange" color when dying. It may grow to 20 in. (50 cm.) but is typically much smaller, particularly when washed up. This specimen was about 3 in. (7.6 cm.) long.

Where to Find It

Subtidal zone, down to 39 ft. (12 m.).

When to Look

Summer.

Notes

This is surely one of the prettiest algae. It is often found washed up into the wrack. Sometimes the midrib is barely visible, or may be missing if you find a fragment. The grainy dots feel like grains of salt. One collector calls it "Pink *Ulva* with measles", reflecting its flat blade, thickness, and bumpy surface.

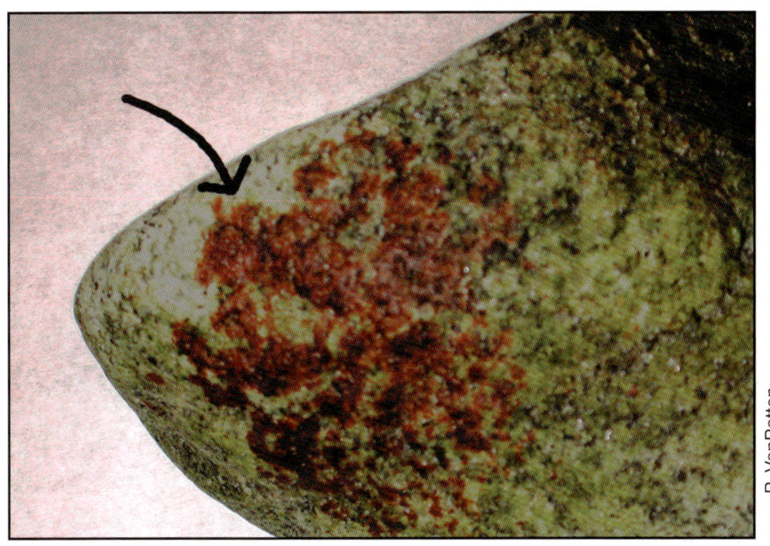

P. VanPatten

Hildenbrandia rubrum
Rusty Rock

What it Looks Like

A thin blood-red or rust colored crust that adheres tightly to rock. The thickness is only half a millimeter thick–about 1/50 of an inch, and it is fairly smooth, so it may resemble thick nail polish or paint.

Where to Find It

Water's edge at low tide, down to 6.5 feet (20 m.) depth.

When to Look

Year-round; when tide is low, on pebbles, cobble, and large rocks at the water's edge in the intertidal zone and beyond.

Notes

Some people seeing this encrusting species for the first time wondered "why someone had painted the rocks red?". There are several other encrusting algae in LIS; this one is very common. This is a great example of a form that most people don't recognize as seaweed. When reproductive, small dots may be visible on the surface with a hand lens. Compare to *Ralfsia* and *Phymatolithon*.

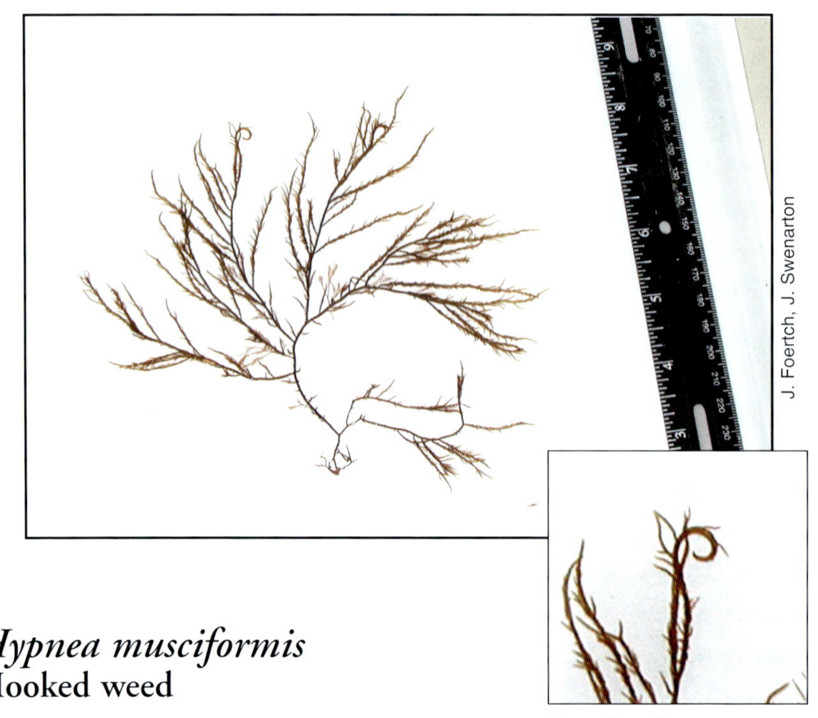

J. Foertch, J. Swenarton

Broad hooks appear on some branch tips.

Hypnea musciformis
Hooked weed

What it Looks Like

Brown color, branching, distinguished by the coiled "hooks" on some branches. Hooks are only on the tips, which is one way to tell it from *Bonnemaisonia.*

Where to Find It

Water's edge at low tide.

When to Look

Year round but infrequent. This species appeared in abundance in eastern LIS in October, 2005.

Notes

Bonnemaisonia is usually a brighter red color, and *Hypnea* more brown, even though it is classified with the red algae. Compare also with another hooked species, *Cystoclonium*. *Hypnea musciformis* contains carrageenan, but is not harvested to the extent that *Chondrus crispus* is, in many parts of the world.

C.R. Janiak

Lomentaria baileyana

What it Looks Like

Profusely branched and rebranched, somewhat rounded ot pyramid shape. Color can be medium red to brownish or purplish red. Typically grows from 3-16 in. (7-42 cm.) tall. Forms short tufts. Looks very much like its relative, *Lomentaria clavellosa*, except that *baileyana's* branchlet tips are pointed, and its older and secondary branches tend to curve more. Of the two, this is the native species.

Where to Find It

Subtidal zone; attached to rocks or other algae; floating in eelgrass beds, or occasionally free-floating. May grow down to about 26 ft. (8 m.).

When to Look

Early summer to early winter.

Notes

This specimen is from a pressed herbarium sheet and is about 3 in. A fresh specimen would be a bit plumper. It is considered a summer annual, but may be able to survive the winter as a small disc-shaped holdfast or a creeping filament.

P. Van Patten

Club-shaped tips on the branchlets are a key feature.

Lomentaria clavellosa

What it Looks Like

Generously branched and re-branched, pyramid shape or round. Soft, straight branches are tapered at both ends, and tips are rounded or club-shaped. Color may vary from rosy red to brownish or purplish red. May be 4-8 in. (10-20 cm.) or more

Where to Find It

Calm waters, subtidal zone. May be attached to other algae or free-floating. This specimen was growing on *Ascophyllum nodosum.* May grow to a depth of about 16 - 39 ft. (5 - 12 m.)..

When to Look

Summer.

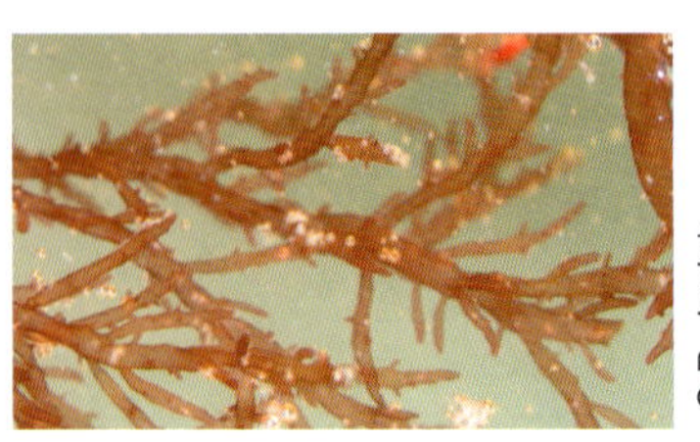

C.R. Janiak

Notes

L. clavellosa is an invasive species, first spotted in New England in Boston Harbor in 1966. Related species: *Lomentaria baileyana* (previous page) is a common native in Long Island Sound, which has pointy branch tips. The species name *clavellosa* comes from a word that means club-shaped. Most abundant in the summer.

C.R. Janiak

Mastocarpus stellatus

Turkish Washcloth

What it Looks Like

Looks much like Irish moss (*Chondrus crispus*). It can be purplish red to dark brown, or may look mottled with pale areas. Branches come off a single stalk base, and edges curl or twist under. An alternate phase in summer is a blackish green rounded crust covered with erect filaments. Usually 2-6 in. (5-15 cm.) long. Branch axes are about the width of a coffee stirrer, or slightly wider.

Where to Find It

Subtidal rocks on the open coast. May be exposed at low tide, probably on the side of the rock facing the ocean. Often above and overlapping into *Chondrus crispus.*

P. Van Patten

When to Look

Year-round; reproductive in fall and winter.

Notes

This species, formerly *Gigartina stellata,* gets its common name from the erect projections on its surface, like terrycloth, when it is in the crust form. *Mastocarpus* branches are curlier than *Chondrus* and reproductive bumps form on the blade surfaces (inset), whereas *Chondrus* has reproductive bumps on blade tips only. The crust stage was first thought to be a separate species, "*Petrocelis*".

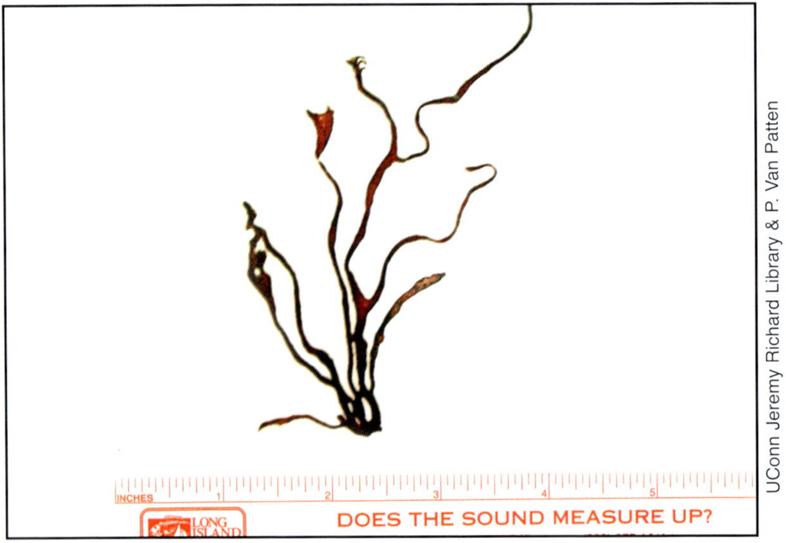

UConn Jeremy Richard Library & P. Van Patten

Nemalion helminthoides
Rubber Threads

What it Looks Like

Brick to purplish red color, slippery, worm-like soft branches. Branches are cylindrical, usually about the thickness of a coffee stirrer in parts, but the width varies. Branches tend to become flatter and thicker at branching points. Usually grows to a height of 2 to 8 in. (5-20 cm.).

Where to Find It

Exposed open coast, mid to low intertidal zone, on rocks.

When to Look

Summer.

Notes

Uncommon. Compare to *Dumontia contorta.*

C. Yarish

Neosiphonia harveyi
Siphon Weed, Polly

What it Looks Like

Slightly coarse, "hairy", light brownish or purplish-red to black color (darkens with age.). Slightly spongy textured, about 1-4 in. (3-10 cm.) tall. Many branches and branchlets, and is attached at the base. Seen microscopically, branches have four cells in a bundle around a central axis, so you can see two rows of cells side by side. (The other two can be seen from the opposite side.) It is not banded like *Ceramium.*

Where to Find It

Open coast, coastal ponds, intertidal to subtidal, growing on rocks, larger algae such as *Codium,* or at the base of salt marsh grass, as in the habitat photo.

When to Look

Late spring to early winter best.

Notes

This very common species was formerly *Polysiphonia* but was recently renamed *Neosiphonia.* See *Polysiphonia* entries.

P. VanPatten

Deep water form, with new individuals sprouting from outer edges

Palmaria palmata
Dulse, Dillisk

What it Looks Like

Large, flat, firm, thick blades, up to 20 in. (50 cm.) tall. From a single blade fanning out from a small discoid holdfast, it divides into sections or lobes like the fingers of a hand, thus the name. Reproductive ones, like the one in the inset, may have many tiny new plants all along the margins of the blades. Color varies from "dusty rose," pink, to light red, to dark red/purple or red-brown in deeper waters.

Where to Find It

Open coast, estuary. Subtidal. Often growing on *Laminaria* stipes.

When to Look

Year-round.

Notes

This alga is collected, rinsed, and dried, then eaten like potato chips or crumbled into salads or soups. It is commercially harvested for food in Maine, and in Europe it is served in snack bowls in pubs. Compare with *Grateloupia turuturu,* an invasive nuisance species.

P. VanPatten

Phycodrys rubens
Sea Oak, Oak Leaf Weed

What it Looks Like

Surprisingly, this species looks like little pink oak leaves. The thin blade sections have both a midrib and veining, (quite unusual for algae). The blades are deeply lobed, thus the resemblance to oak leaves. Blade edges are somewhat serrated ("toothed"). The color is quite striking, usually pale pink but may vary to bright pink, and, in deeper waters, burgundy. Generally less than than 8 in. (20 cm.) long, often much smaller, as in the photo.

Where to Find It

Subtidal zone to 33-49 ft. (10-15 m.) depth, but may wash up in drift. May form "meadows" in the shallow subtidal zone.

When to Look

Year-round.

Notes

This is arguably one of the most attractive red algae to be found in the Sound. The edges were beginning to dry a bit when this photo was taken. A pressed specimen can be framed as art.

P. VanPatten

Phyllophora pseudoceranoides

Close-up of a tip shows reproductive growth occuring along length of branch, not only tips.

What it Looks Like

Long, thin, main axis is dark brown and cylindrical, resembling a young tree twig, branches widen out abruptly at the tips into bright red flat blades. Looks like little red ribbons on a twig. This species is capable of growing quite large (nearly 20 in., 50 cm. or more) but a foot or less is more typical for the Sound.

Where to Find It

Subtidal, on boulders or in areas with shifting sands, also kelp holdfasts.

When to Look

Year-round.

Notes

Closely resembles *Coccotylus truncatus* but becomes vegetatively reproductive on various parts of the blades, whereas *C. truncatus* becomes reproductive only at the blade tips (see p. 62 to compare). *P. pseudoceranoides* is often habitat for marine bryozoans.

C. Yarish

Conn. College Arboretum

Any hard surface is good habitat for algae. Here, a specimen growing on *Ascophyllum.*

Phymatolithon laevigatum
Rock Plant

What it Looks Like

Pinkish to purple thin, flat, crust with white edges, adhering to rocks or other hard surfaces.

Where to Find It

Rocky intertidal, tightly stuck to the rocks or on coarse seaweeds.

When to Look

Year-round.

Notes

Compare this encrusting-type seaweed to *Hildenbrandia* and *Ralfsia.*

Symmetrical branching pattern, highly magnified

Plumaria plumosa
Red Feathers

What it Looks Like

Delicate, very small, highly branched, “feathery”; light to bright pink or purple-pink. Magnification reveals very symmetrical opposite branching, in a flat plane. About 2-8 in. (5-20 cm.) tall.

Where to Find It

Subtidal, deep waters, attached to larger algae; sometimes found beneath a canopy of *Ascophyllum,* or on base of *Laminaria.*

When to Look

Year-round; spring and summer best Uncommon but worth hunting.

Notes

The bright purple-pink color is easy to spot if it washes ashore, despite its small size. This alga is named after plumes because of its resemblance to soft showy feathers. Synonym: *Plumaria elegans.*

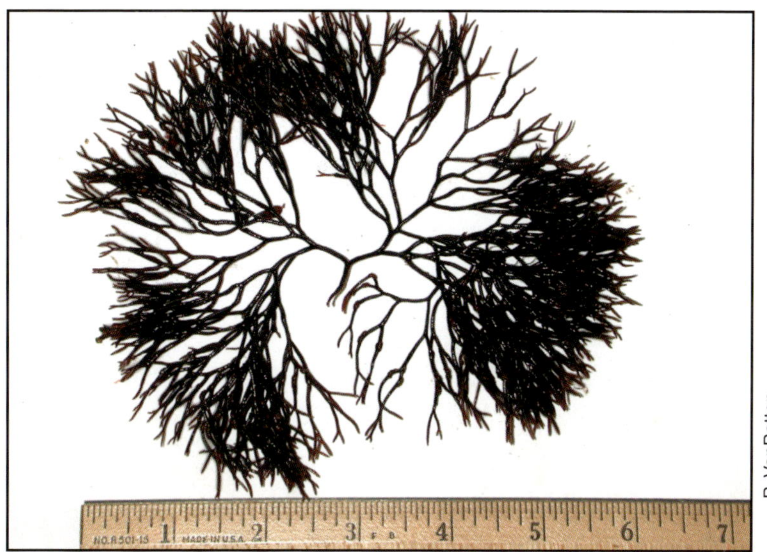

P. VanPatten

Polyides rotundus
Twig Weed

What it Looks Like

Dark brown to black color; rounded shape. Somewhat "elastic" branches are cylindrical and uniform thickness, about the thickness of a lollipop stick, or a bit thinner if young. Branching is very symmetrical, and dichotomous (Y-shaped). It may branch 6-8 times or so, ending in y-shaped forks. The tips are sometimes white or creamy colored. Size may be 3-8 in. (8-21 cm.)

Where to Find It

Open coast, estuaries, tide pools and low subtidal zone; kelp holdfasts.

When to Look

Year-round.

Notes

This red alga is often mistaken for a brown because of its dark hue. It prefers deep waters. *Polyides* contains some carrageenan, but is not commercially harvested.

J. Foertch, J. Swenarton

Polysiphonia lanosa
Polly

What it Looks Like

Small (about 1-1.5 in. or 3-4 cm.) dark tufts, looking somewhat wiry or stiff, with forked branching and rebranching; nearly always growing on *Ascophyllum,* as above. Color is dark purple to black. The many *Polysiphonia* species have cylindrical branches with cells grouped around the center like bundles of sticks or bones. The number and orientation of these cells is used to distinguish one from another.

Where to Find It

Open coast, intertidal to subtidal, nearly always growing on *Ascophyllum.*

When to Look

Year-round.

J. Foertch, J. Swenarton

20-24 tightly bundled cells around a center are visible under a microscope.

Notes

This species is common in the eastern Sound. Viewed under a microscope, has 20-24 cells tightly bundled around the central axis. Related species: *P. fucoides, P. nigra, P. subtilissima,* and many more Pollys! *See also Neosiphonia.*

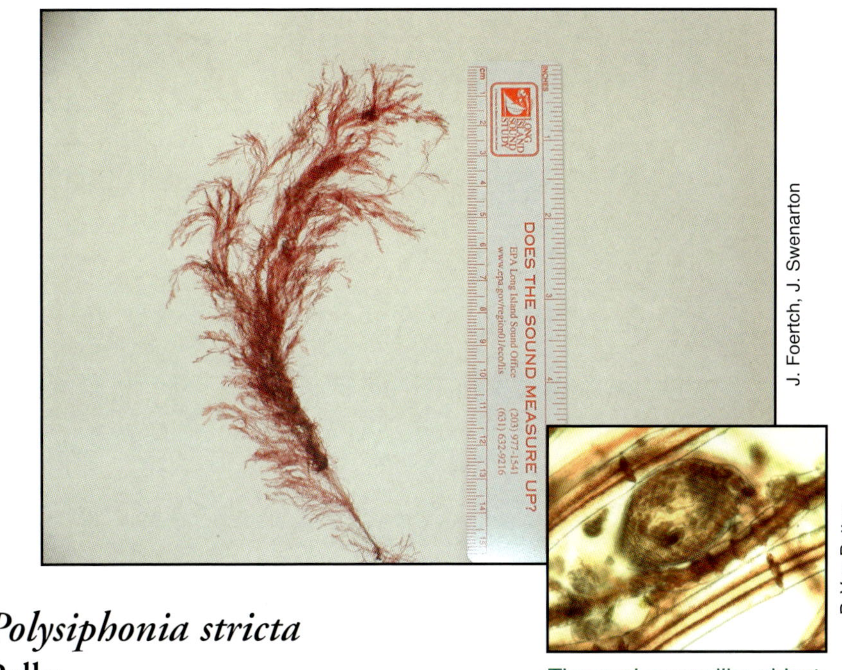

J. Foertch, J. Swenarton

P. Van Patten

The oval or urn-like object in the center is a female reproductive structure.

Polysiphonia stricta
Polly

What it Looks Like

Looks hairy, a little coarse; nearly always growing on a larger alga. Color is dull red to burgundy wine, or nearly black. May grow up to nearly 12 in. (30 cm.) "Pollys" are hard to identify without a microscope, but the habitat helps. The many *Polysiphonia* species have cylindrical branches with cells grouped around a center like bundles of sticks or bones. The number and orientation of these cells is one way to distinguish them. *P. stricta* has 4 cells around a center, so you see 2 in the inset (the other 2 are on the other side). Microscopic urn-shaped pericarps (female reproductive structures) are another distinguishing feature.

Where to Find It

Open coast, cool waters of the intertidal to subtidal zones, growing on *Ascophyllum*, or sometimes *Fucus*.

When to Look

Late winter to spring.

Notes

Related species: *P. fucoides, P. lanosa, P. nigra, P. subtilissima,* and many more "Pollys" not included here. *See also Neosiphonia.*

C. Yarish

C. Yarish

Polysiphonia subtilissima cells.

Polysiphonia subtilissima
Polly

What it Looks Like

This "polly" may be reddish green, blackish purple, or quite black, and is often found at the base of *Spartina* grass at the water's edge of salt marshes. It can tolerate nearly fresh water. Its soft, bushy filaments form dense mats. See more about *Polysiphonia* in the descriptions of *P. stricta* and *P. lanosa.* Grows to nearly 6 in. (15 cm.).

Where to Find It

Brackish or muddy areas, shallow warm coves, intertidal to subtidal zone, often growing on other algae. May be in salt marsh ditches or at the base of salt marsh grass. May also be attached to shells or rock.

When to Look

Year-round.

Notes

We might call this the "Pigpen" of the "Pollys", as this Polly is attracted to mud. When viewed under a microscope, has 4 cells grouped around the central axis (see inset). There are many species of *Polysiphonia*, so a small sample is presented here.

UConn Jeremy Richard Library

P. VanPatten

Polysiphonia nigra
Polly, Brown Polly

What it Looks Like

Small and delicate tufts, like fine hair. Many are brown colored, but colors may vary to more reddish purple, to red-black. Somewhat less densely branched than most other "Pollys". May grow up to almost 8 in. (20 cm.)

Where to Find It

Open coast, low intertidal to upper subtidal zone, attached to rocks, sometimes other algae.

When to Look

Year-round.

Notes

Viewed under a microscope, *P. nigra* has 12 to 18 cells grouped around the central axis. It differs from the other "pollys" in that the group of cells twist to form a spiral pattern. This is obviously a pressed specimen.

C. Yarish

R. Wikes

Porphyra leucosticta
Laver, Nori

What it Looks Like

Roundish, elongated, or oblong shape. A flat, very thin, membranous blade; no branching. Color is typically reddish brown in LIS but varies, so may be more purplish or greenish. Feels slippery like plastic food wrap but more fragile; only one cell thick. 4-6 in. (10-15 cm.) long or more.

Where to Find It

Open coast, subtidal, attached to other larger algae.

When to Look

Spring and summer.

Notes

Nori is commercially cultivated elsewhere for use in making sushi wrappers, and for its pigments, which have medical applications. Related species: *P. linearis; P. umbicalis,* and other *Porphyra.* When rounded, *P. leucosticta* is nearly impossible to distinguish from *P. umbicalis* without a microscope and a specialist handy, or a DNA analysis.

C. Yarish

Porphyra purpurea cells, greatly magnified. the blade is only one-cell thick, and feels like wet clinging food wrap.

Porphyra purpurea
Laver, Nori

What it Looks Like

Roundish, elongated, or oblong shape. A flat, tissue-thin, membranous, slippery blade; no branching. Color is typically reddish brown in LIS but varies, so may have more red, greenish, or purple.

Where to Find It

Open coast, subtidal, attached to other algaesuch as *Ascophyllum.*

When to Look

Spring and summer.

Notes

Male and female reproductive parts occupy opposite sides of the same specimen, but staying distinctly separated as though there is a zipper down the center. The top edge, with raggedy white spots, is the male side. Related species: *P. linearis* (not shown but looks quite similar in shape and color) and other *Porphyra. See also Notes on other Porphyra entries.*

P. VanPatten

C. Yarish

Reproductive structure.

Porphyra umbicalis
Laver, Nori

What it Looks and Feels Like

Roundish, elongated, or oblong shape. A flat, very thin blade; no branching. Color is typically reddish brown in LIS but varies, may look more olive-red. Feels slippery like plastic cling wrap, and is about the same thickness, but is more fragile. Size is typically 4-12 in. (10-30 cm.) high.

Where to Find It

Open coast, upper intertidal, attached to rock or, more often, other algae.

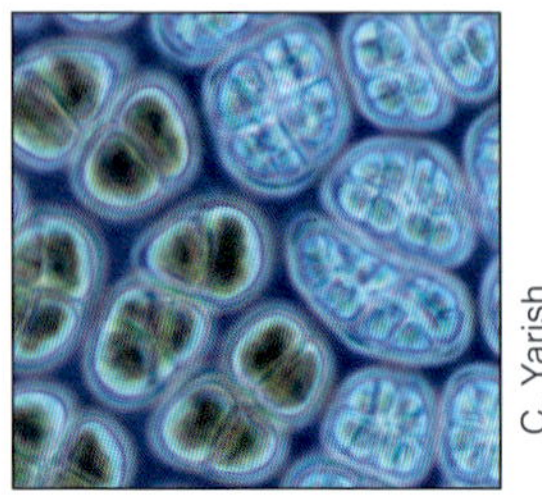

C. Yarish

Cells from surface of *P. umbicalis, as seen under a microscope.*

When to Look

Spring and summer.

Notes

Like other *Porphryra*, you can eat this species, dried and flaked into soup or with fish, etc. A similar species is commercially cultivated for making sushi wrappers, and for its pigments, which have medical applications. Related species: *P. linearis; P. miniata; P. leucosticta.* It is best left to a specialist to distinguish them.

S. Taylor, Connecticut College Arboretum

Ptilota serrata
Red Fern

What it Looks Like

Red-brown to dark brown color. Highly branched, slightly stiff, with many fortified small branchlets resembling small spines or thorns (but not sharp) projecting from the larger branches coming from the base. Although branching is mostly opposite, pairs may differ (one side shorter than the other). Grows to nearly 6 in. (about 15 cm.) tall.

Where to Find It

Open coast, cool, deep waters, subtidal as low as 12-15 ft.(3.6-4.5 m.); may be attached to rocks or the base of large subtidal brown algae, such as kelp.

When to Look

Year-round, but cool months best.

Notes

Not common; prefers cooler waters north of Cape Cod, but is occasionally sighted in Long Island Sound. Its range may be moving northward, as in the case of *Callophyllis cristata.* It has been recently seen in the Noank, Connecticut area.

P. VanPatten

Rhodomela confervoides
Tufted Red Seaweed

What it Looks Like

Densely branched, hair-like main branches with short tufted branchlets at the tips. If one uses imagination, it could resemble a bouquet of small flowers. Main axes are about the width of a paper clip. Color is dark red to brown or purplish-brown, with tips possibly brighter red. The color shown here is typical. May grow as large as about 15 in. (38 cm.).

Rhodomela's female reproductive structures, under the microscope. Top, female "pericarps".; Bottom, male branches with bumpy tips carrying "spermatangia".

Where to Find It

Subtidal; usually attached to rocks, but may be floating if it detaches and washes up.

When to Look

Spring to fall.

Notes

This alga's structure is similar to *Polysiphonia.* The short, flat-topped clusters at the branchlet tips are the best clue. Densely tufted branch tip portions may fall off in late summer, making identification harder.

P. Van Patten

J. Sears, NEAS

Spermothamnion repens

Red Puff Balls, Red Tufts

What it Looks Like

Small, dense, round, very soft filamentous tufts, up to about an inch and a half (3.8 cm.) in diameter; color pinkish red to bright red.

Where to Find It

Low intertidal to subtidal; attached to hard surfaces or growing on other algae; may detach and float freely or accumulate in masses. May be found with *Corallina*, as in the inset photo (white species on the right).

When to Look

Year-round.

Notes

This species has caused alarm in Rhode Island by detaching in masses and washing onto public bathing beaches during swimming season. Like most, this alga is completely harmless, unrelated to toxic "red tide". Harmful red tides are caused by microscopic algae that emit a toxin. When any mass of algae is decaying in great clumps, naturally there will be an odor, but it shouldn't scare anyone away from the beach! In such an event, why not collect it, and celebrate its beauty by making a pressed specimen, as described in the Introduction. These were found in calm waters at Groton, CT in July, after several days of rain.

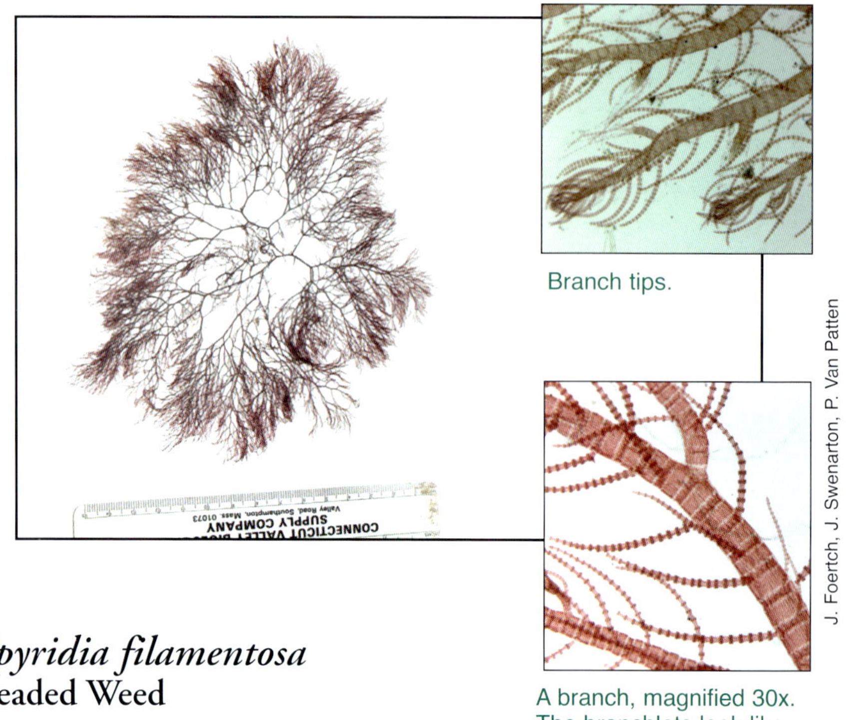

Branch tips.

A branch, magnified 30x. The branchlets look like small beads.

J. Foertch, J. Swenarton, P. Van Patten

Spyridia filamentosa
Beaded Weed

What it Looks Like

Very small (less than 8 in., or 20 cm.) red filamentous tufts, gracefully curved. You'll need a microscope to tell this from similar species. Its central axis branches are thicker than its many incurved branchlets, which are arranged opppositely or in whorls. Under the microscope you can see that the branches and branchlets are divided into nodes by rings of tiny cells, giving it the appearance of tiny beads on a string.

Where to Find It

Protected areas, warm coves. Subtidal, or washed ashore, generally attached to shells or rock, or other algae.

When to Look

Early summer.

Notes

This is a charming annual, warm water species, found only in the summer in Long Island Sound. It may bleach to a pale color. It may become luxuriant in very protected warm places.

Simplified Key to Some Common LIS Seaweeds

A "key" is a tool that can be used to identify organisms. Like a detective, you make careful observations about your specimen, then select from the choices below the description that best fits. Then follow the numbers to the logical conclusion, the name you are seeking. Next, compare it to the picture and description in this guide. **Note:** f your specimen doesn't fit any of the listed species, check the more comprehensive taxonomic keys listed in the bibliography. This key was adapted, updated, and expanded from *A Wader's Guide to the Shore* by Sally Taylor.

1. Seaweed is colored bright green, yellow-green, blue-green to dark green, cells stain purple-black with iodine.. **Green Algae**
1. Seaweed is olive green, brown, red, purple, or black 2
2. Specimen is olive green, yellowish-brown, tan, or medium brown, no or very little red.. **Brown Algae**
2. Specimen is otherwise.. 3
3. Specimen is pink, red, purple-red, brown-red, or nearly black* .... **Red Algae**

* It can be very difficult to tell brown and red algae apart. One test is to put a fragment in boiling water. If the red pigment comes out, it is likely red, not brown.

Green Algae - Chlorophyta

1. Plant is large, dense, dark green and spongy, forking branches aare thick and cylindrical ....*Codium fragile*
1. Plant is small and bright green, hair-like, forming carpets on rocks...*Blidingia minima*
1. Plant otherwise...2
2. Plant is bright green or yellow-green, a tube or flat sheet.. 3
2. Plant otherwise...4
3. Plant is a flat sheet, with or without holes ....*Ulva lactuca*
3. Plant is a thin inflated tube, shaped like an intestine ....*Ulva intestinalis*
4. Plant has long, branching filaments that mat together, looks cloudy under water...*Cladophora* sp.
4. Plant branching but not as above...5
5. Plant is light or dark green, feathery, mossy or fern-like fronds...*Bryopsis* sp.
5. Plant is in small round tufts, soft or stiff, with rope like branches ....*Acrosiphonia arcta*

continued on next page

Brown Algae - Phaeophyta

1. Plant is large, coarse and leathery, long and narrow, or U-shaped, no midrib 2
1. Plant otherwise..3
2. Plant is a flat, thick, single blade with stipe and claw-like holdfast, up to several feet or more in length, divided on one end into segments like fingers on a hand...*Laminaria digitata*
2. As above, but without blade divisions, with hollow stipe........................ ....*Saccharina longicruris*
2. As above but without blade divisions and solid stipe, ruffled edges .... *Saccharina latissima*
2. Plant is a flat blade but without large claw-like holdfast............................3
3. Plant is a flat blade, paper-thin, light brown, no stipe, solitary or in a group.. *Petalonia fascia*
3. Plant is a flat blade, medium to dark brown and thick, may have small granular dots on surface but no stipe.......................................*Punctaria spp.*
3. Plant otherwise..4
4. Plant is rounded, globular but lumpy like cauliflower or brains hollow inside, yellow to brown...*Leathesia difformis*
4. Plant otherwise..5
5. Plant is a dark crust on rock or shells..................................*Ralfsia verrucosa*
5. Plant is soft, finely branched, hairlike, growing on other algae, pilings, or rock, looks "cloudy" underwater...................................... *Ectocarpus silicosus*
5. Plant is large (1 ft. or more), draped on intertidal rocks, branching, with forked tips that may be swollen..6
5. Form is cord-like, dark brown, branched or not...7
6. Branches inro forks, blades has a midrib and paired float bladders along blade.. *Fucus vesiculosus* (rockweed)
6. Plant has firm, single float bladders along main axis, olive green or olive-brown color...*Ascophyllum nododum*
6. Plant otherwise..7
7. Plant is a single, unbranched, long brown cylindrical filament of rather uniform diameter, with or without small hairs along the length.................8
7. Plant otherwise..9
8. Plant is single, smooth, long unbranched cord or tube, tapering at one or one end, with small disc holdfast ....*Chorda filum*
8. Plant as above but with many very short fuzzy hairs all along its length ....*Halosiphon tormentosus*
9. Plant is dark brown and cylindrical, but with many curving side branches from one main axis...*Chordaria flagelliformis*

continued on next page

9. Plant single filaments, tubes with frequent constrictions, like a string of sausage links ....*Scytosiphon lomentaria*

Red Algae

1 Plant a flat sheet, crust, or blade..2
1. Plant branched 4
2. Plant a rust or deep red crust on rock.........................*Hildenbrandia rubrum*
2. Plant a flat sheet...3
3. Plant tissue-thin, slippery, color brownish-red, attached to rocks or other algae in intertidal zone ....*Porphyra* spp.
3. Blade thin and semi-transparent, color light to bright pink, slight midrib, has granular dots on surface..*Grinellia americanus*
3. Blade thick, like heavy paper or leather, large, divided into sections from a single base, like fingers on a hand.................................. *Palmaria palmata*
4. Small wedge-shaped red blades, at branch tips, attached to a twig-like cylindrical axis...9
4. Plant small, rigid, like bones or rock, branched into symmetrical articulated joints, fan-shaped.. *Corallina officinalis*
4. Plant otherwise...5
5. Plant shrub-like, densely branched with main axis fanning out from a narrower base, rounded branch tips divide into two's, branches in one flat plane...*Chondrus crispus*
5. Plant small, finely branched...6
6. Christmas-tree shape, branches with some tiny, distinctly coiled hooks along branches, not only at tips....................................*Bonnemaisonia haminifera*
6. Plant bushy, highly branched with hook-shaped coils at tips of some branches only... *Hypnea musciformis*
6. Plant without hooks...7
7. Red to brown, branches have distinct dark banding pattern; branch tips shaped like pincers or crab claws..*Ceramium spp*
7. Hooks and pincer-tips absent................................ 8
8. Plant filaments resemble a string of tiny barrels or tub-shaped beads.......... ....*Champia parvula*
8. Other than above, red-brown to purple or black, much-branched, "hairy", often growing on *Fucus, Ascophyllum* or *Chondrus;* viewed microscopically has long, narrow cells bundled around a central axis............*Polysiphonia spp.*
9. Reproductive buds occur at tips of blades only.............. *Coccotylus truncatus*
9. Reproductive budding occurs along parts of blade other than the tips....... .... *Phyllophora pseudoceranoides*

**but see also the invasive species Grateloupia, not included here. Grateloupia looks similar but is very slippery to the touch and has many irregular shapes.*

Bibliography - More suggested reading about the algae

Basic Biology

H.C. Bold and M.J. Wynne. 1985. *Introduction to the Algae*, second edition. 720 pp. Basic reference book for serious students of the algae. Prentice-Hall.

M. Dring. 1982. *Biology of Marine Plants.* 199 pp. Edward Arnold.

L. E. Graham and L.W. Wilcox. 2000. *Algae.* A marvelous up-to-date, user-friendly textbook. 640 pages plus glossary. Prentice-Hall.

Taxonomy - Keys for the serious seaweed specialists

M. Villalard-Bohnsack. 1995. 145 pp. *Illustrated Key to the Seaweed of New England.* Rhode Island Natural History Survey. Rhode Island Natural History Survey. A popular, lavishly illustrated key.

J. Sears, ed. *NEAS Keys to the Benthic Marine Algae of the Northeastern Coast of North America, from Long Island Sound to the Strait of Belle Isle.* Northeast Algal Society (NEAS) and Connecticut Sea Grant. 161 pp. A very comprehensive key with broader geographical range than the previous entry.

W.R. Taylor. 1962. *Marine Algae of the Northeastern Coast of North America.* 2d edition. Univ. of Michigan Press. 509 pp. Once the "bible" of Northeast algal taxonomy but now out of print. It is the most complete volume, but many items are outdated so use in conjunction with the newer keys.

Ecology and Distribution

K. Lüning, 1990. (English translation by C. Yarish) 1990. *Seaweeds: their Environment, Biogeography, and Ecophysiology*

Economic Uses:

C. Lembi and J.R. Waaland, eds. 1988. *Algae and Human Affairs.* Discusses origins of algae and commercial uses in food, industry, agriculture, medicine, and more.

C. Yarish, C. Penniman and P. Van Patten, eds. 1988. *Economically Important Marine Plants of the Atlantic.* 158 pp. Connecticut Sea Grant.

Cooking and Eating Algae

Cooksley, Valerie G., R.N. 2007. *Seaweed:Nature's Secret to Balancing Your Metabolism, Fighting Disease, and Revitalizing Body and Soul.* Sttewart, Tabori, and Chang. New York. 204pp.

Ellis, Lesley. 1998. *Seaweed: A Cook's Guide.* Easy-to-prepare, simple and tasty recipes. Available in paperback. Fisher Books. 100pp.

Gusman, J. 2003. *Vegetables from the Sea.* Harper Collins Publishers. 131pp. Great recipes from a noted gourmet cook. Lavish illustrations, hardcover. 128 pp.

Maderia, Crystal J. 2007. *The New Seaweed Cookbook.* North Atlantic Books. 155pp. Recipes of all sorts, some from author's grandmother.

Pressing Algae

The Guide to Pressing Seaweed by Alex Frost and Molly Fallon. The Cryptogamic Botany Company, Peacedale, R.I. Paperback. 32 pp. Gives detailed instructions with many photos to perfect your pressing technique.

Index

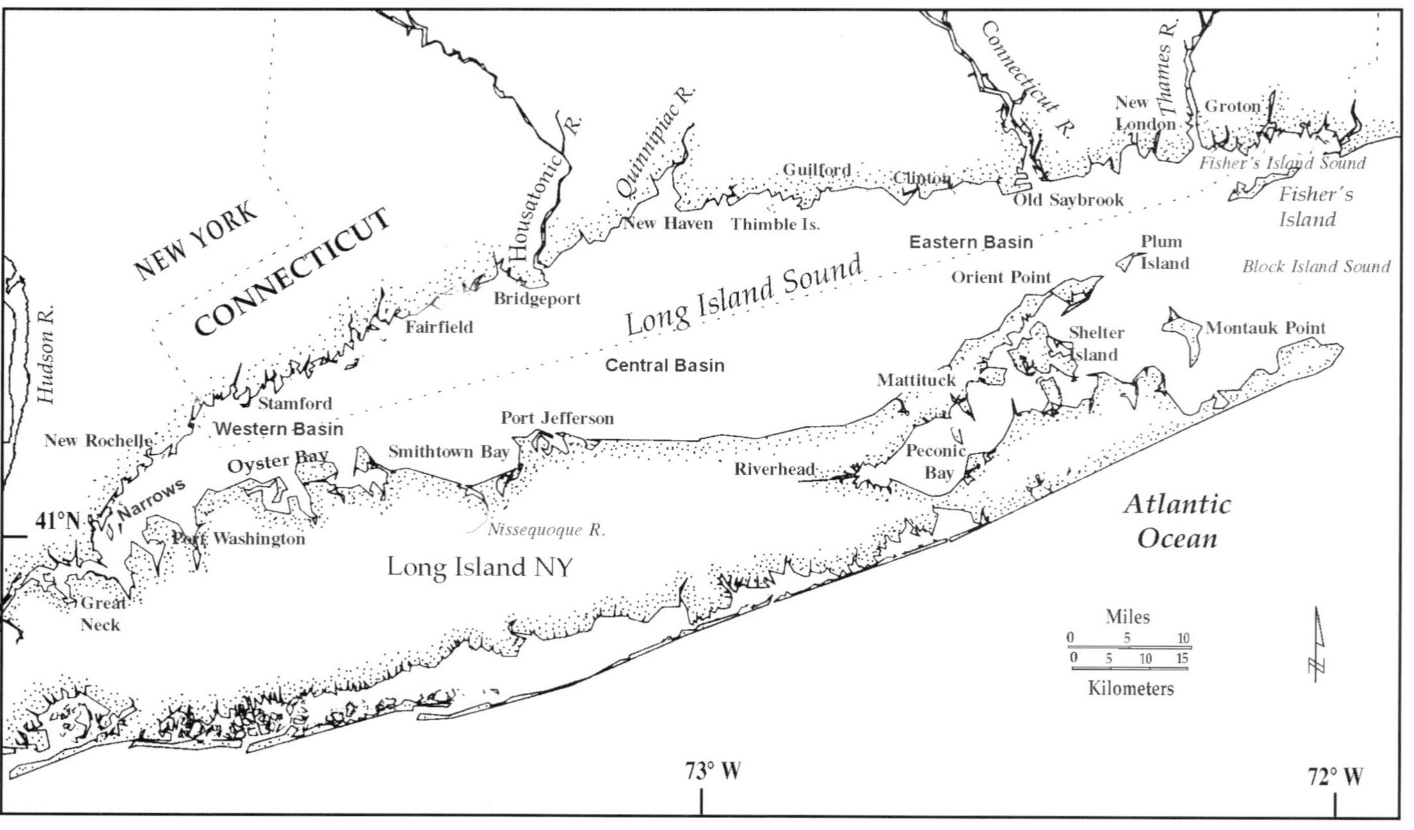

LONG ISLAND SOUND U.S.A.